THE
VEGETABLE
GARDEN
PEST
HANDBOOK

Quarto.com

© 2021 Quarto Publishing Group USA Inc.
Text and Photography © 2021 Susan Mulvihill

First Published in 2021 by Cool Springs Press, an imprint of The Quarto Group,
100 Cummings Center, Suite 265-D, Beverly, MA 01915, USA.
T (978) 282-9590 F (978) 283-2742

Cool Springs Press titles are also available at discount for retail, wholesale, promotional, and bulk purchase. For details, contact the Special Sales Manager by email at specialsales@quarto.com or by mail at The Quarto Group, Attn: Special Sales Manager, 100 Cummings Center, Suite 265-D, Beverly, MA 01915, USA.

25 24 23   5

ISBN: 978-0-7603-7006-3

Digital edition published in 2021
eISBN: 978-0-7603-7007-0

Library of Congress Control Number: 2020949442

Design and illustration: Mattie Wells
Cover Images: Shutterstock
Photography: Susan Mulvihill, except: Alamy: page 67 (right); Gary Bernon, USDA APHIS, Bugwood.org: pages 117 (left), 201 (right column, top); Clemson University, USDA Cooperative Extension Series, Bugwood.org: pages 67 (left), 195 (left column, top); Whitney Cranshaw, Colorado State University, Bugwood.org: pages 64, 65, 73, 79 (right), 83 (right), 91, 102, 103, 194 (right column, top two), 195 (right column, second down), 197 (right column, second from top), 199 (left column, middle & second from bottom); David Jones, University of Georgia, Bugwood.org: pages 66, 194 (right column, bottom); Ken Gray Photograph Collection (P 256), Special Collections and Archives Research Center, Oregon State University Libraries: pages 72, 195 (right column, top); Bill Mulvihill: pages 182, 185 (bottom right), 186 (bottom), 202; Eugene E. Nelson, Bugwood.org: pages 77 (left), 196 (left column, top); Russ Ottens, University of Georgia, Bugwood.org: pages 90, 197 (right column, top); Lucy Potts: pages 93, 109 (left), 197 (right column, middle), 200 (left column, bottom); Marcia Sands: pages 86, 197 (left column, middle); Shutterstock: pages 69, 77 (right), 78, 79 (left), 95 (right), 110, 111, 114, 115, 116, 121, 123 (right), 127 (bottom), 129 (top), 195 (left column, second from top), 196 (left column, middle & second from bottom), 198 (left column, top & second from bottom), 199 (left & right columns, bottom), 200 (right column, bottom), 201 (left column, bottom three); Ronald Smith, Auburn University, Bugwood.org: page 65; Alton N. Sparks, Jr., University of Georgia, Bugwood.org: pages 83 (left), 196 (right column, second from bottom); Sharon Watson: pages 76, 92, 196 (left column, second down), 197 (right column, second from bottom)

Printed in China

THE VEGETABLE GARDEN PEST HANDBOOK

Identify and Solve Common Pest Problems on Edible Plants

SUSAN MULVIHILL

COOL
SPRINGS
PRESS

This praying mantis is ready to help you control troublesome insect pests.

CONTENTS

INTRODUCTION

Growing vegetables is a passion of mine. My love affair with them began the moment my grandmother took me by the hand to show me the wonders of her garden. Soon, I was raising my own zucchinis and tomatoes. I was hooked.

Throughout my adult life, I've understood the importance of knowing how to grow your own food. It's a skill everyone should have and is simpler than you might think. Gardeners everywhere know how nutritious and delicious homegrown produce is, and that's what keeps them heading back to the soil year after year.

But there's one frustrating challenge many gardeners would rather not deal with: bugs!

In this book, you will find tools to identify vegetable garden pests so you can make informed gardening decisions. Here's a sampling of the resources that await you:

> Do some sleuthing in the Vegetable Crops and Potential Pest Problems table (page 36). Find the crop you're growing, the damage you've spotted, and the likely culprit.

> Learn important details about the most common vegetable pests: what they look like, their life cycle, the type of damage they cause, and ways to control them.

> Identify beneficial species (page 124) so you can easily recognize them and perhaps watch them in action.

Once you've pinpointed the problem pest, chapter three points you toward the most successful organic products and methods. You'll even receive step-by-step information for implementing many of them with DIY projects (page 148).

Have you come across a bug in your garden that you don't recognize? Just scan the mugshots beginning on page 194 to identify it, then find details about it in chapter two. You'll also find many useful references and sources for the organic products I've mentioned.

Whether you are a beginning gardener or have been growing your own food for years, I think you'll agree that each season is a learning experience. While some years are a bit more challenging than others, making new discoveries is how we improve our skills.

As a Master Gardener and seasoned garden writer, I help individuals with specific pest challenges and post more generally on social media about how to deal with garden pests, particularly those that want to eat our veggies. I wrote this book to collect those scattered pieces as a single resource you can return to again and again to accompany you on your gardening journey. My goal is to provide useful information for all gardeners without being so technical your eyes glaze over.

Here you can learn how to thwart pests before they even arrive, partner with beneficials, and keep your plants so healthy, they can survive or even fend off inevitable attacks. My focus is on natural, low-impact (to the environment, not the pests!) controls for pests that become problematic. The most important lesson I've learned over many decades of gardening is that you *can* grow a bountiful garden and outsmart the bugs without the use of synthetic chemicals. I am so excited to share this with you. Let's get started!

Healthy vegetable gardens should include flowers to attract pollinators and beneficial insects.

A bountiful harvest is the goal of vegetable gardeners everywhere.

INTRODUCTION TO ORGANIC PEST MANAGEMENT IN THE VEGETABLE GARDEN

1

Growing your own food is one of life's greatest pleasures. The act of nurturing young seedlings, being out in the fresh air, harvesting that first vine-ripened tomato, and knowing you are putting healthy food on the table all combine to make it such a positive experience.

At least, it is until that first time you head into the garden and discover holes in the broccoli leaves or the biggest caterpillar you've ever seen nibbling on your tomato crop. Those aren't exactly positive experiences, are they?

Whether you are a seasoned gardener or a beginner, you know insects and other bugs, such as spider mites, will be one aspect of growing a garden. And that isn't necessarily a bad thing. All kinds of bugs—both the good and the bad—play important roles in a healthy environment.

Before we proceed, a clarification is in order. In this book, I'll discuss insects, spiders and spider mites (which are arachnids), pillbugs and sowbugs (crustaceans), and slugs and snails (mollusks). Within the animal kingdom, insects, arachnids, and crustaceans are all classes of arthropods. To be an arthropod, an animal must have a segmented body, jointed appendages, and an exoskeleton, and also lack a backbone (invertebrate). That's pretty confusing, isn't it? My goal is to keep the information simple and relatable. Since we all refer to these

creatures as bugs and insects, that is exactly what I'm going to do in this text!

Humans have identified approximately 1 million insect species on our planet. If that freaks you out, take comfort in knowing that only about 1 percent of those species are pests to us. That puts things in a whole new perspective, doesn't it? Or, at least, I hope it does.

The remaining species—approximately 990,000—are either beneficial or benign. Beneficials—bugs that include lacewings, ladybugs, ground beetles, and yes, even spiders—spend their days munching on aphids, insect eggs, cutworms, and so many other problematic pests. Other beneficials pollinate the flowers of edible or ornamental plants. You'll meet these and many more on page 124. Get to know them as your partners in creating a healthy ecosystem in your garden.

Benign species, while they aren't particularly helpful in controlling pests, don't bother humans or the crops we grow in our gardens or on farms. Many of them perform a critical role, feeding on decomposing plant material and building and aerating the soil in the process.

Myriad species of bugs contribute to the balance of our planet's various, complex ecosystems. In our gardens, beneficials keep the pests from taking over and making the act of growing our food just about impossible, the benigns help keep the soil alive, and the pests feed the beneficials. If you let them, and even help them by how you garden, beneficials and benigns can extend that environmental balance to your garden by managing pests and improving your soil.

As you read this book and learn to embrace and enhance your environment in order to be a successful gardener, it is my hope you will also discover that the world of bugs is really cool. This guide will increase your awareness of what's out there, help you figure out which ones might be causing problems, and choose the most environmentally friendly ways to resolve those problems. We gardeners have plenty of options!

WHAT IS ORGANIC GARDENING AND WHY IS IT SO IMPORTANT?

We've all heard the term "organic gardening," yet it often has different interpretations. Put simply, it means gardening without the use of chemicals. In this context, chemicals are synthetic fertilizers, herbicides, or insecticides (pesticides) made in a lab to mimic or try to improve upon those found in nature.

But organic gardening also involves maintaining soil health, regularly monitoring the garden for potential problems, and understanding and using the role insects and other bugs play in the environment. The many organic vegetable gardens I have visited in the US and Europe demonstrated to me that this method can be both beautiful and productive. What I observed has pushed me to embrace organic methods in my own garden for many years with excellent results. If you are interested in growing healthy produce for yourself, your family, neighbors, and community, chemicals need not play a role in the process.

FERTILIZERS

Synthetic (inorganic) fertilizers provide a quick fix for struggling plants and lawns but kill the microorganisms that make nutrients available to the roots of plants. Their nitrogen and phosphorous levels typically exceed the needs of the plants to which they are applied. Irrigation and rainfall wash that excess away to contaminate our rivers, lakes, and ground water. In contrast, organic fertilizers contain lower but more diverse concentrations of nutrients. Plants more readily absorb them and the nutrients improve the soil.

HERBICIDES

While weeding isn't a whole lot of fun, using synthetic chemicals to get rid of them can do more harm in the long run. Manufactured herbicides often last longer on plant material and in the soil than we realize, affecting the quality and safety of our gardens. Planting in soil treated with a broad-spectrum weed killer can make your new plants sick, and the residual herbicide may kill them. I like to use grass clippings to mulch my vegetable garden. However, if I used herbicides on my lawn—to kill dandelions or other broadleaf weeds—those lawn clippings could potentially wipe out the majority of the plants. Why? Most vegetable plants are broadleaf plants!

INSECTICIDES

For a long time, we gardeners have been encouraged to spray for bugs "just in case," meaning whether or not those bugs are present. But insecticides are some of the most problematic chemicals—for both our food and the environment—used in gardening.

United States law requires the label of every insecticide, organic or synthetic, to include that product's ingredients and recommended safety precautions—how to apply it, whether it is safe around children and pets, how long to wait before harvesting and eating the produce protected by the insecticide, how to dispose of the product, and so on. Having this knowledge is for the consumer's protection, yet I worry about the potential effects of long-term use.

Growing a healthy, productive garden is a joyful experience.

Help keep beneficial insects, including pollinators such as bees, safe by avoiding the use of chemical pesticides.

Insecticides are short-term solutions. Some pests in a given population will survive being poisoned, living to pass their resistance to their offspring. Even if you kill all the insects in your garden, more will arrive to take up residence. But pest species are not the only ones you should consider.

Many pesticides are non-selective poisons that kill pests, benign species, and beneficial species indiscriminately. Some of these species lost to "friendly fire" may have even solved your pest problem for you. Even birds and other animals that eat pests can suffer as the insecticides make their food toxic or kill so much of it there's not enough left. In the escalating war on insects, chemical companies have developed neonicotinoids, the newest "insecticide" that is decimating both native and introduced pollinators. Without our pollinators to help us grow food, life on this planet would be precarious at best.

Killing off all of the insects is *not* the secret to having a healthy, productive garden! We should strive to achieve a balance instead. I grow vegetables organically because I want to know that they were grown and handled in a healthier way. It's very reassuring to know synthetic chemicals didn't play a role in their production. Let's embrace the natural balance of our gardens when pesticides and other chemicals aren't part of the equation.

INTEGRATED PEST MANAGEMENT (IPM)

Integrated Pest Management, or IPM, is a systematic approach of choosing the most environmentally friendly method of handling pest problems.

The first step involves regularly monitoring your garden. The pest profiles in chapter two will remind you of this frequently because it's a simple way to successfully to control pests. By making a routine of walking through your garden, you can observe how well your plants are growing and spot problems right away. I like to go out into the garden with a cup of coffee in hand and casually stroll around to see how everything is doing. It's a simple and pleasurable daily routine.

If you notice a new critter on a plant, take the time to identify it correctly. In the next section, I list a few ways to accomplish this. You might discover your new visitor is a beneficial, in which case you can celebrate that it's chosen your garden for a home base.

When your research shows that a pest has arrived, determine if you need to take action and at what point. Maybe you'll just need to pick them off the plants. It depends on what your tolerance is for damage on some of the crops you are growing. We humans like to have pristine, perfect produce, but imperfect food can be just as tasty.

If the problem appears to be escalating, create a plan and put it into action. This might involve some type of mechanical control, such as a barrier to keep the pests off plants; one or more cultural practices to help the plants tolerate pest damage; or biological controls, such as introducing natural predators.

If you're familiar with IPM, you know the final control option revolves around the use of synthetic chemical pesticides, although this is always done as a last resort. I suggest following an organic IPM program by eliminating the chemical controls altogether since they create more problems than they resolve.

No matter which action you take, evaluate how well your plan worked and document it for the future. This thoughtful approach to pest issues will increase your gardening knowledge and enjoyment, as well as your success.

BUG IDENTIFICATION 101

I'll be the first to admit that some bugs are downright alarming to look it—especially when they've been highly magnified under a microscope. But we absolutely must not judge them by appearance alone. While many of them might look harmful, they usually are doing good deeds for us in our gardens. Remember they are an important part of the ecosystem. Fight the urge to squish a creature based on your gut reaction alone. Wouldn't you feel badly if you killed a bug, only to later learn that it gorges on aphids?

I know I'm repeating myself, but the most important step is to properly identify a bug before deciding if you need to take action. There are many useful resources to help you do this.

I think all gardeners should have a hand lens. These inexpensive magnifying tools allow us to check out the details of a bug's body, such as the number of legs it has, the color of its head, or perhaps some unusual antennae. For bugs that won't hold still, catch them in a butterfly net or a small jar so you can get a good look.

If you discover a pest that I did not include in this book, the internet is an easy resource for attempting to ID it. Enter a basic description, such as "large black-and-white striped beetle." Include the state or region in which you found it, or you'll be overwhelmed by hundreds if not thousands of images, and most won't be relevant. I can't tell you how many times I've nailed the identification by using this method, and in a short period of time, too.

Searching for "insect identification keys" on the web can point you to tools that step you through questions about the creature's appearance and perhaps where you found it. This might guide you to a positive identification or, at the very least, let you know to which order or family the bug belongs. That's always a good start.

A hand lens makes identification easier by magnifying the details.

Have your smartphone with you? There's an app—several, in fact—for identifying these creatures. I've found Picture Insect – Bug Identifier, Insect Identification, Leps by Field Guide (for butterflies and moths), and Seek by iNaturalist to be useful.

Horticultural programs at colleges and universities are another resource that will often assist with identification. Talk to the staff at nurseries, garden centers, and public gardens; they are well-trained and knowledgeable. You will find additional information in Resources on page 190.

Remember that the majority of insects go through different stages during metamorphosis into an adult. Most start out as an egg, hatch into a larva (perhaps a grub or a caterpillar), pupate (for example, butterflies do this in a chrysalis, moths in a cocoon), and later emerge as an adult. This means you might not recognize an insect in a different stage than the one you know.

WHAT'S YOUR "BUG" TOLERANCE?

I've got a question for you: When you walk into your garden and see one or two problem insects, do you dive into action or wait to see if you'll need to intervene? In some cases, particularly when it comes to vegetable gardens, it's important to jump in right away while the pests are at a vulnerable stage and few in number. But remember that beneficials can help us out of a tight spot.

A few years ago, I was walking through my garden and stopped to check on the currant bushes. When I saw a lot of puckered leaves, I inwardly groaned. Aphids. I turned over some leaves. Yes, lots and lots of them were on the undersides. I debated about what I wanted to do but decided I didn't have the time at the moment.

You guessed it: My schedule got busy over the next few days and I completely forgot about those aphids. A few days later, I suddenly remembered them and dashed back out to the garden. Instead of witnessing more aphid mayhem, I was pleasantly surprised to discover that almost all of them had vanished. In their place was a collection of adult ladybugs, their larvae, and even some pupae, which are essentially pre-adults.

As you can see in the photo opposite, ladybug larvae look rather frightening. Fortunately, I knew exactly what they were and that their voracious appetite for aphids is legendary. They had come to the rescue and enjoyed quite a feast, courtesy of my forgetting about the problem.

The moral of the story is this: If we are patient, beneficials will often take care of problems for us. Chapter two profiles the most common vegetable garden pests and the timing for using simple organic controls—and it indicates the natural predators that target each pest.

Ladybug larvae look very different from the familiar adults we all know and love.

FOLLOW GOOD CULTURAL PRACTICES

What would you say if I told you there are many simple steps you can take, with a minimal investment in time and money, to ensure your garden grows well? Such steps are called cultural practices. A gardener who takes a holistic approach to all aspects of growing a garden organically becomes proactive rather than reactive. Instead of taking steps in *response* to problems, the gardener prevents the problems or greatly reduces the chance that they will occur.

Research studies show that stressed plants send out volatile signals that, unfortunately, attract harmful insects that are only too happy to attack a plant with weakened defenses. Healthy, vigorous plants fare much better, defending themselves against pests, sustaining minimal damage, and recovering more quickly. If your plants are in good shape, you will have more time to spot the problem and intervene.

Following the practices outlined here, you can grow a healthy, productive garden.

CHOOSE THE BEST LOCATION.

Vegetable plants thrive if you provide them with the most ideal setting possible. Nearly all crops require a minimum of six hours of sunlight daily. When considering locations, keep an eye on how the sun moves through your yard and how that changes at different times of the growing season. Where do buildings, fences, or large trees cast shadows? Try to steer clear of those areas if it's an option.

The good news is that there are ways to get around these challenges. Think about growing climbing vegetables on an arbor or trellis. The higher they grow, the more sunlight they can potentially reach. Climbing vegetables include cucumbers; pole beans; peas; vining summer squash, such as trombone zucchini; as well as small melons and winter squash.

'Musica' pole beans are my personal favorite for vertical gardening. I grow them on an arbor that spans the pathway between two raised beds. In addition to being a very cool feature of the garden each year, the plants get plenty of sunlight, the beans hang down on the inside, and I get to stand in the shade while harvesting them. I heartily recommend this!

If the only sunny area of your yard is a deck, patio, or balcony, no problem: Grow your vegetables in containers. You can even place them on casters and move them around to capture the maximum amount of sunlight.

Did you know many vegetable crops are tolerant of a little shade? Look at your options in the list below, but remember that the general rule of thumb is to choose the sunniest spot available.

SHADE-TOLERANT VEGETABLES

- Beet
- Bok choi
- Broccoli
- Cabbage
- Carrot
- Celery
- Chard
- Chinese cabbage
- Kale
- Kohlrabi
- Leek
- Lettuce
- Mizuna
- Mustard
- Pak choi
- Scallion
- Spinach
- Turnip

And, last but not least, consider a north/south orientation for your beds. That way, as the sun moves across the sky, plants won't cast shadows on their neighbors as much as they would in an east/west orientation.

Avoid areas where roots from trees or shrubs infiltrate the soil because they will compete with your vegetable plants for water and nutrients. If that is your only choice, don't worry: I have seen a beautiful and productive garden that is planted in large pots and unusual containers. Rocks in the soil are another challenge, especially when growing deep-rooted crops such as carrots.

Drainage is an important consideration, as well. Water tends to pool in low spots, but vegetable plants won't thrive if their roots are always wet. Aim for a level planting area if you can, and consider adding compost to heavy soils that drain slowly.

Raised beds are an excellent choice if your soil doesn't drain well or is too rocky. Just build a bed, fill it with soil, and you've got a garden. Learn how to make one on page 170.

KNOW YOUR REGION'S GROWING CONDITIONS.

By being aware of when your area's last and first frosts typically occur, you can plant your vegetables at the appropriate time.

Many cool-season crops—beets, carrots, onions, parsnips and spinach, for example—are tolerant of chilly temperatures and prefer growing either early or late in the season. If planted during the peak temperatures of summer, they will struggle, and you're going to have a disappointingly short harvest—if you get one at all. Conversely, if you jump the gun by planting warm-season crops, such as eggplants and tomatoes, before the danger of frost has passed and your area has a cold snap, the plants will likely be toast. Even if they survive, being exposed to cold temperatures is a horrible way to start off their lives in the garden.

If you are new to the area or to gardening, talk with garden center staff, local garden clubs, or gardening friends to learn about your region's growing season. This is a good way to meet your gardening neighbors! Chapter four lists additional resources for locating research-based information. Then you'll be able to choose the right time for planting vegetable crops.

TAKE CARE OF YOUR SOIL.

It's ironic that we gardeners often ignore the one thing on which our gardening success most depends: the soil! After all, it is literally the foundation of a garden. We completely forget about taking care of the soil because we are so excited to start planting.

If you take good care of your soil, it will in turn take good care of your plants.

Look closely at a scoop of soil (use your hand lens!). You'll be amazed by the amount of beneficial life it contains: earthworms, springtails, beetle larvae, centipedes, and more. If you have access to a microscope, you'll see a host of microbes including bacteria, fungi, nematodes, and protozoa. While some of those might sound like bad critters, rest assured that the activities of the vast majority of them build healthy soil structure and make nutrients available to plant roots.

Because of this, we should disturb the soil as little as possible. Traditionally, we gardeners were taught to rototill the soil each spring before planting. While it might seem like a great idea to create light, fluffy soil, it actually destroys the structure that insects and microorganisms worked so hard to create. This includes tiny air pockets necessary for soil-borne creatures and plant roots to thrive. Tilling can also bring dormant weed seeds to the surface, where they will sprout.

Consider taking a "no-till" approach. Instead of turning over your soil with a rototiller or shovel, apply organic amendments to the surface of the soil instead. They will filter down into it on their own. The best organic amendment to use is compost. I apply a 1- to 2-inch (2.5 to 5 cm) layer of it to the top of the soil in each raised bed in the spring and fall.

There is one exception to the no-till approach, however. Some pests will lay eggs or pupate in the soil, then hatch or emerge in the spring. It is acceptable to disturb the top couple of inches of the soil in the fall to expose the eggs or pupae to predators and cold, wet weather conditions.

Grow vegetables vertically to save space in your garden and capture more sunlight.

WATER EFFICIENTLY.

Water is a precious resource on our planet. Yet vegetable gardens require regular watering to grow and produce well. As gardeners, our goal must be to take steps to conserve water while helping our gardens thrive. While that might sound like a contradiction, it's possible to do both at the same time.

Plants become stressed if they are consistently under- or overwatered throughout the season. Since stressed plants send out chemical signals that attract pests, we want to avoid that sequence of events, right?

It can be difficult to find time in our schedules to water our gardens regularly, but there are solutions. Consider setting a water timer so your garden is watered at consistent intervals. Even if you don't have an automated sprinkler system, basic water timers that run on batteries are available at home and garden centers and are easy to program. Just hook one up to your water spigot and your garden will get the water it needs while you're taking care of other things.

Keep in mind that the same amount of watering time won't apply to the entire growing season. As plants grow and the weather warms up, they require additional moisture. Monitor their appearance and adjust accordingly.

It is pretty easy to recognize signs of underwatering: The plants will wilt and appear unhealthy. Do note, however, that during extremely hot weather, even a well-watered plant may wilt to minimize its exposed surface area and therefore water loss.

It is a little harder to spot overwatering problems. If you have to dig up a plant to figure out why it's failing to grow properly, you might discover the roots have been rotting—a sign that the plant was getting too much water. To prevent either of these extremes, poke your finger into the soil an inch or two on a regular basis to see how moist or dry it is, and adjust your watering routine as needed.

Early morning is the best time of day to water vegetable plants. This puts moisture in the root zone so plants can draw upon it during the day as they transpire (breathe) and grow in the sunlight. Early in the season, watering in the evening will chill the soil and plants overnight. During hot summers, it might be necessary to water both in the morning and later in the day to provide enough moisture for your plants.

The most efficient way to water is through the use of either a drip irrigation system or soaker hoses. Both deliver water directly to the soil at the base of the plants, right where it's needed most.

If you can, avoid using overhead sprinklers. This inefficient watering method loses a lot of moisture to evaporation. Also, having water or soil splash onto leaves can spread plant diseases. Tomato plants are particularly susceptible to soil-borne pathogens spread in this manner.

To conserve moisture, put a layer of organic mulch on the soil surface. Ideal materials include shredded leaves, weed-free straw, or grass clippings from a lawn that has not been treated with herbicides. Apply mulch at a thickness of 1 to 3 inches (2.5 to 7.6 cm), but do not mound it around the base of the plant; this will invite rot, disease, and pests.

As you read through the pest profiles in chapter two, you'll learn that a thick layer of mulch can prevent certain pests from laying their eggs in the soil beneath their host plants. That's a pretty ingenious—and simple—strategy to employ.

CHOOSE RESISTANT VEGETABLE VARIETIES.

If you have had problems repeatedly with a specific pest on a crop, do a little research to find out if there are cultivars (varieties) that are resistant to that pest. For example, 'Flyaway' carrot is resistant to the carrot rust fly. There are also bean and squash cultivars that are resistant to Mexican bean beetles and squash bugs, respectively. It's always fun trying something new, and if the pests won't bother them, that's even better.

GIVE YOUR PLANTS SOME ROOM.

When you are ready to plant your seeds and seedlings, space them appropriately so they will have plenty of room to develop. Look for spacing guidelines on seed packets, the tags that came with your plants, or in a gardening book. Your plants will reward you with healthy growth and good productivity.

When plants are crowded together, it is easier for pests to hide in and infiltrate the entire crop. One year, I planted a crop of lettuce quite thickly. Since I knew slugs could be a problem, I sprinkled organic slug bait on the soil surface. To my dismay, I started seeing holes in the leaves and the damage kept getting worse. I soon realized that by growing the lettuce so closely together, those slugs didn't even need to slither down to the ground in order to move on to the next plant. I had essentially created an aerial slug highway for them. Had I spaced the seedlings appropriately, it would have

A garden journal is an invaluable tool for growing a garden successfully.

been harder for the slugs to move from plant to plant. Why simplify things for these annoying pests?

Some plants require vertical support—such as a sturdy stake, cage, or trellis—to keep them from sprawling on the ground. In addition to providing better air circulation for the developing vegetables or fruits, that support makes them less accessible to certain insects and critters.

KEEP A GARDEN JOURNAL.

One of the simplest, most useful cultural practices in which you can engage is recording how your garden performs each year. A journal doesn't have to be fancy—just a notebook will do—but what's inside will be worth its weight in gold to you.

The purpose of keeping a journal is to learn from our gardening successes and failures. I don't know about you, but I always think I'll remember to do something differently next year. What typically happens, however, is that life gets in the way and before I realize it, that great idea is gone forever.

In your journal, you can jot down the weather conditions, the dates of your last spring frost and first fall frost, when you first planted peas or tomatoes in the garden, something new you learned, or a technique you want to try the following year. Perhaps you heard about a great new cultivar that excelled in a friend's garden under challenging conditions. Be sure to note the pests you encounter in your garden. You might record when you first spotted aphids or tried pest control solutions that worked or didn't work.

Occasionally, there will be a year when a particular type of pest wreaks havoc in your garden. In 2017, a lot of my gardening comrades and I had terrible problems with pillbugs and sowbugs. Ordinarily, they are quite innocuous in the garden, but not that year. Those little terrestrial crustaceans chewed through the succulent stems of our young melon and cucumber seedlings.

And in 2019, it was the "year of the earwigs" and, once again, I discovered I wasn't alone. They caused so much damage. Noting experiences such as this in a garden journal provides us with valuable information and solutions for the future.

One thing that makes gardening easier for me is listing when to do certain tasks. If I have written them down, I don't have to waste time trying to figure it out all over again each year.

No matter what type of information we include, all of these things are noteworthy. Writing them down—and, of course, reading them again throughout the growing season—makes us much better gardeners.

PRACTICE CROP ROTATION.

You are going to hear a lot about crop rotation in this book. Why? It is a useful strategy for preventing or reducing pest problems.

You've probably heard this term before and dismissed it, thinking that crop rotation is just for large-scale farmers. And while it might be challenging for those with tiny gardens, let me explain why it's important.

Crop rotation refers to changing the location of where you grow families of crops each year. If you grow the same type of crop in the same location year after year, you might encounter these two problems:

1. Different types of crops utilize different amounts of nutrients from the soil in order to grow and produce. If you continue growing a specific crop in the same spot without replenishing those nutrients for the following year's garden, the plants will deplete the soil and subsequent plantings will underperform.

2. Even more importantly, specific pest or disease problems afflict certain crop families. Those pests or disease pathogens can remain in the soil where the crops were grown.

Since I've already talked about the importance of taking good care of your soil, let's focus on the second problem, because it is particularly applicable within the context of this book. You will be able to use your knowledge of crop rotation to your advantage to prevent or lessen problems with recurring pests. Interested?

The first thing you'll want to do is familiarize yourself with the families of commonly grown vegetable crops.

Knowing your vegetable plant families is so important and useful. By being aware that some pests and diseases are specific to a plant family—and rotating your families of crops—you will have the most success at growing vegetables. This is a case where seeing the big picture makes all the difference!

PLANT FAMILY NAME	VEGETABLES WITHIN THIS FAMILY
Asparagus (*Asparagaceae*)	Asparagus
Beet (*Amaranthaceae*)	Beet, spinach, Swiss chard
Cabbage or Mustard (*Brassicaceae*)	Arugula, broccoli, Brussels sprouts, cabbage, cauliflower, kale, kohlrabi, mustard, radish, rutabaga, turnip
Carrot (*Apiaceae*)	Carrot, celery, chervil, cilantro, dill, fennel, parsley, parsnip
Cucurbit (*Cucurbitaceae*)	Cucumber, melon, pumpkin, squash (summer and winter)
Grass (*Poaceae*)	Corn
Legume (*Fabaceae*)	Bean, pea
Nightshade (*Solanaceae*)	Eggplant, pepper, potato, tomatillo, tomato
Onion (*Amaryllidaceae*)	Chives, garlic, leek, onion, shallot
Aster or Sunflower (*Asteraceae*)	Artichoke, endive, lettuce, sunflower, tarragon

Don't let weeds get out of control in your garden.

For example, carrot weevils (*Listronotus oregonensis*) cause a significant amount of damage to carrot family crops, most notably carrots, celery, and parsley. They don't bother other plant families so it makes sense to use crop rotation in your garden to thwart them.

Vulnerable to fusarium and verticillium wilt, early and late blight, and mosaic virus, members of the nightshade family benefit greatly from crop rotation. Crops in the cucurbit family are also good candidates since they can be hosts to some nasty pests. Other plant families benefit from being moved around in the garden each year as well.

Record your garden layout every year. Many years ago, I created a template of our garden beds on my computer. Each year, I print out a fresh copy, list what I'll be growing in each bed, and store those records in a notebook for future reference. If your gardening journal has a pocket or is a binder, that's an ideal place to keep everything together.

Prior to each garden season, I pull out the previous three years' records. Plants in the nightshade family are my first focus. I select beds where I've not grown any of these crops over that time. Then I go through the same routine with each family, from the most vulnerable to the relatively pest-free veggies, to create the current year's layout. While it takes a little effort, it's been worth the time because I have minimal pest and disease problems in the garden.

What if you have a small garden? You can still practice crop rotation using these methods:

1. At the very least, move nightshade family crops to different areas of the garden each year as they tend to be the most prone to serious plant diseases.

2. Alternate growing your nightshade family crops in your garden space one year and in containers the next year or two. At the end of each growing season, always discard the potting soil from your containers as disease pathogens might be present.

3. If you're not growing those types of crops but have had pest problems with other plant families, use steps 1 and 2 for better success in growing them.

4. If certain pests or plant diseases are a recurring issue for you, look for cultivars of plants or seeds labeled with resistance to those obstacles.

Crop rotation is one of your best tools for growing a healthy garden organically. In addition, being tuned in to potential pest or disease problems gives you a leg up on what preventative measures to take. You will learn much more about this in chapter two.

KEEP UP WITH YOUR WEEDING.

It's a fact that weeding is not on my list of fun things to do (and probably not on yours either). But there are many good reasons I need to stay on top of this chore.

Weeding slows me down long enough in different areas of the garden so I can check how my plants are doing. I also know weeds compete with my plants for moisture and nutrients, and I don't like that at all. It stresses them and invites damaging pests to the garden. In addition, weeding is a good way to find—and kill—cutworms since they hide in the top inch of soil during the day.

Here are the strategies I've learned for making it difficult for weeds to grow. By following them, I have more time to do the fun things in the garden. And the best part is they don't involve any chemicals.

1. **Use weed-block fabric in pathways.** In my large raised-bed garden, the paths are covered with weed-block fabric (a.k.a., landscape fabric), which in turn is covered with a few inches of bark mulch. Because the fabric is heavy duty and tightly woven, it's extremely difficult for weeds to sprout through it and grow. With this method, I only have to weed the top of each raised bed, which is a tremendous time saver. If any weeds do pop up in a path, they are really easy to pull out of the loose mulch.

   While we're on the subject of using weed-block fabrics, I don't recommend them for use within landscape beds. While they are permeable, they can interfere with the flow of water through the soil, which makes plants struggle. They are very helpful for permanent pathways, however, and I prefer to leave it at that.

2. **Don't stir up the soil in your beds.** I used to drag a cultivator across the surface of the soil during the growing season, just to make it look prettier and fluffier. I didn't realize I was actually bringing hundreds of weed seeds to the surface where they were only too happy to germinate. As soon as I realized I was making more work for myself, I discontinued the practice.

3. **Mulch your beds.** I place a thick layer of mulch on the surface of most of my beds for two reasons: It impedes weed growth and helps the soil retain moisture longer. Good mulching materials include shredded leaves, weed-free straw, and grass clippings from a lawn that hasn't been treated with herbicides.

4. **Consider using plastic mulch for warm-season crops.** Its main purpose is to increase the soil temperature, which is what warm-season crops, such as tomatoes, melons, cucumbers, peppers, and eggplants, need. This is especially helpful in my short growing season. In addition, having that cover over the soil surface makes it very difficult for weeds to grow. You'll also learn in chapters two and three that reflective mulches make it difficult for some pests to find their host plants.

CLEAN UP GARDEN DEBRIS.

When it comes to dealing with pests, keeping your garden free from debris is one of your most important cultural practices.

Many insects and other bugs overwinter or lay eggs in crop residue. By leaving it in place rather than tidying it up at the end of the season, you are just about guaranteed to have the same problem the following year. For example, if carrot rust flies (see page 72) found your carrot patch last year, their larvae or pupae overwintered in the roots and will emerge this year, thus perpetuating the problem. Instead of giving up on the crop and leaving it in the soil to decompose, pull up all affected roots by the end of the season to eliminate or greatly reduce future issues.

As you work your way through the pest profiles in chapter two, you will see that cleaning up garden debris is a straightforward way to eliminate or reduce pest problems. Remember to dispose of any plant material that is likely to contain eggs or other stages of pests; those plants are not good candidates for the compost pile.

The same applies to diseased plants. Remove them from your garden as quickly as possible since they will attract pests in addition to spreading pathogens. Throw them in the garbage to prevent those diseases from returning next year—do **not** compost any diseased garden waste.

Put out the welcome mat for pollinators and other beneficials by planting many types of flowers, shrubs, trees, and grasses.

ATTRACT POLLINATORS AND OTHER BENEFICIALS TO YOUR GARDEN

You might think that *attracting* insects and other bugs to your garden is the last thing you'd ever want to do. On the contrary: A garden that is home to diverse species should be the goal of all gardeners.

In our gardens, bugs quietly do great things for us without our knowledge. When we slow down and look closely, we'll see plenty of good guys, with some being very familiar and others looking a little scary. You'll have the opportunity to learn more about specific types of beneficials beginning on page 124.

It's easy to feel like the world is filled with bugs that are out to get our plants and crops. But as I mentioned earlier, of all the insects in the world, only one percent of them are damaging. Those remaining are either beneficials (such as pollinators, predators, or parasitoids) or benign, which do no harm. If you're wondering what parasitoids are, they live on or inside the body of a host and eventually kill it—a bug's version of the movie *Alien*!

For all of those annoying aphids in the garden, plenty of good critters are more than happy to feast on them: assassin bugs, big-eyed bugs, earwigs, lacewings, ladybugs, and spiders, just to name a few.

Think about all of the wonderful pollinators, such as honeybees and native bees, wasps, hoverflies, butterflies, and moths. If you are growing a vegetable garden (or tree fruits or berries), you *want* those pollinators close at hand.

There are three simple—and enjoyable—ways to attract both pollinators and beneficials: Create a diverse landscape, plant flowers within your vegetable garden, and/or build an insect hotel. Let's take a look at each.

LANDSCAPE FOR YOUR HELPERS.

Many beneficials require nectar and pollen in order to survive. Plenty of predatory beneficials don't consume either but will be attracted to, and thrive in, a diverse landscape. Provide all of them with a variety of sizes and types of trees, shrubs, flowers, and grasses. Avoid planting a monoculture, such as a single species of flower. Most gardeners probably don't mind the suggestion to select a wide variety of plants, right?

One popular way of attracting beneficials is by creating "insectaries," which are permanent habitats that meet their needs within a landscape. The plantings, which supply pollen and nectar, usually include perennials, annuals, wildflowers, and native bunchgrasses.

When choosing flowering plants, make sure they'll provide season-long bloom, and match them to your conditions. For example, if you locate your insectary in a dry area of your landscape, select drought-tolerant plants.

Different types of bugs have different mouthparts. Certain types of flower heads will attract different creatures seeking nectar and/or pollen. By selecting plants with a variety of flower heads, you'll get the best return on your investment in terms of insect diversity.

Some excellent choices are asters, bee balm, black-eyed Susan, members of the carrot family (including angelica, chervil, and dill), coreopsis, early-blooming basket of gold, goldenrod, oregano, and sunflowers. Wouldn't it be a treat to have an insectary brimming with these attractive flowers?

Creating an insectary near your vegetable garden will attract natural predators that patrol the area for problems.

This cleverly designed insect "lodge" attracts solitary bees and other beneficials.

INTERPLANT IN THE VEGETABLE GARDEN.

Attract the good bugs by interplanting annual herbs and flowers within the vegetable beds themselves. As beautiful as healthy vegetable plants are, incorporating flowers among them will bring color and helpful bug activity to the area. Mixed plantings also hinder the movement of problem pests from one crop to the next.

Some of my favorite additions are cosmos, marigolds, nasturtiums, sunflowers, and zinnias. Each of these is an annual: They grow, bloom, set seed, and die in a single season. Plant them in the spring after the danger of frost has passed. They require minimal attention on your part other than taking the time to enjoy their beauty. The beneficials attracted to those flowers will also spot pests that are causing problems for your vegetables.

Consider planting cover crops any time one of your veggie beds is vacant. Also known as "green manures," they improve your soil's fertility when you turn them under once they've finished growing. Earlier, I said to avoid disturbing the soil structure. Cover crops are an exception to this approach. By turning over the top couple of inches (5 cm), we will accelerate decomposition of the foliage and roots that in turn will add much-needed nutrients.

Many types of cover crops are nitrogen-fixing legumes that make nitrogen available to subsequent plantings. Where do beneficials fit into this? Cover crops, such as buckwheat, hairy vetch, clover, or peas, provide pollen and nectar for them. My favorite is buckwheat because it grows and flowers in a very short time.

Cover crops planted in the spring are usually turned under by fall. Those planted in the summer will decompose through the fall and winter months, then are incorporated into the soil the following spring.

BUILD AN INSECT HOTEL.

If you've never heard of an insect hotel before, you are in for a treat. They provide a sheltered place for beneficials and pollinators to hibernate and/or lay their eggs. You will find DIY instructions on how your make your own insect hotel on page 165.

My husband and I saw our first insect hotel in a Swiss botanical garden and were intrigued. It was essentially a wooden box containing compartments for insects. Some sections had wooden blocks with holes drilled into them to provide nesting places for solitary bees. Others contained materials attractive to certain types of creatures, such as ladybugs and spiders. (Yes, you want to welcome spiders! Remember they are excellent predators and shouldn't be feared or scorned.) This insect hotel even had a green roof where succulents were growing!

We have built three insect hotels for our garden so far and really enjoy watching the activity in and around them. These types of structures are also referred to as bug hotels, condos, or stacks.

Insect hotels can be any size you want and will provide an interesting focal point for your garden. They're also great educational tools to teach kids (and adults) about the important role insects and other bugs play in the environment and that humans shouldn't step on or otherwise annihilate them.

Insect hotels attract beetles, lacewings, ladybugs, moths, solitary bees, and spiders, among many other creatures. Solitary bees include carpenter bees, digger bees, mason bees, and sweat bees. They do not form hives or colonies like honeybees and bumblebees do.

In our garden, we get a lot of mason bees, which are fascinating to watch. In the spring, the males hatch first. About two weeks later, the females hatch and mate. They look for holes in plant stems or woody plants in which to lay their eggs. The female lays an egg, leaves pollen grains for the larvae to eat after they hatch, seals the chamber with a bit of mud, lays the next egg, adds pollen, and so on. They typically lay multiple eggs in each tube or cavity. The solitary queen lays female eggs at the back of the tube and male eggs toward the front to guarantee the males will hatch first.

Rest assured that mason bees are quite docile and will only sting if harassed. Once you build an insect hotel, you'll soon discover that watching it is both fascinating and cheap entertainment. We always have a couple of chairs nearby so we can sit and observe the different species that use ours throughout the year.

There are only two guidelines for building an insect hotel: It needs a roof to shelter the materials and inhabitants from rain and snow, and the structure should face south or east to capture plenty of early morning light.

My advice is to keep things simple. To get ideas, do a web search on "insect hotels" and prepare to be astounded at the variety of intriguing structures people all around the world have built.

Most of the time, solitary bees will find your insect hotel without any urging. After female bees mate, they cruise around looking for just the right spot to build their new nest. If you do not see much activity after one year, you can purchase mason bee cocoons and introduce them into your garden. Look for examples of suppliers on page 192.

ATTRACT BIRDS TO YOUR GARDEN

There are many excellent reasons why you would want to attract birds to your garden. They are delightful to observe, especially while feeding, bathing, or raising their young. They are a sign of a healthy ecosystem. And, many of them eat pests. Because of this, they can be your partners for pest control. Birds will come to your garden if you meet their primary needs of food, water, shelter, and a place to raise their young.

Many types of birds eat a variety of foods including seeds, suet, berries, insects, and other bugs. Bluebirds, chickadees, finches, kingbirds, nuthatches, phoebes, sparrows, swallows, warblers, woodpeckers, and wrens are all insect eaters, although most will eat other foods, as well. Any of these would be a joy to have in your garden and the more the merrier.

I enjoy watching chickadees hunting for caterpillars. These little birds are quite tame and it is amazing to witness how many times they head for their nest box with a worm dangling from their beaks! If you're thinking you only want to attract the insect eaters, know that it's good to welcome the seed and berry eaters, too. I've found that birds are quite social: The more bird activity you have going on in your garden, the more other types of birds are drawn to it. And what a treat it is to watch them!

By hanging feeders that hold sunflower or thistle seeds, you will attract them very quickly. Chickadees, crossbills, and finches (including goldfinches) adore seed. Some birds change their diet depending on the time of the season. For example, many will switch from a seed-based diet to insects and other bugs to feed their young.

In addition to being lovely to look at, bluebirds are fantastic insect eaters.

Suet cakes hung from tree branches or feeder poles attract nuthatches, pine siskins, thrushes, woodpeckers, and wrens, just to name a few.

Berry-eating birds include finches, grosbeaks, warblers, and waxwings. Many wonderful native shrubs produce berries and look terrific in the landscape. Some bird favorites are American cranberry bush (*Viburnum trilobum*), golden currant (*Ribes aureum*), red-twig dogwood (*Cornus sericea*), serviceberry or saskatoon (*Amelanchier*), and snowberry (*Symphoricarpos*). In addition to their berries, these shrubs have beautiful fall foliage and lovely flowers that attract many pollinators.

Water is key to attracting birds because they need to drink and bathe daily in order to stay healthy. There are many ways to add water to your landscape, from small birdbaths and simple ponds up to large water features. The sound of splashing water will bring them in and it's a treat watching them take baths.

Providing shelter for your bird friends is so important. They need to feel safe from predators and require areas where they can get out of the weather or roost at night. The easiest way to accomplish this is by adding shrubs and trees of varying heights that they can easily fly into at a moment's notice. Planting trees and shrubs near your garden will greatly add to their sense of security.

To offer places for birds to raise their young, you need to know their requirements. Many birds nest in trees, shrubs, or on the ground. Others—including bluebirds, chickadees, nuthatches, swallows, swifts, woodpeckers, and wrens—are more than happy to raise a family in a birdhouse. When building or purchasing a birdhouse, know the dimensions and entrance size that attract the species you are after. You can find a website with these specifications listed on page 191.

As desirable as birds are, I will admit to occasionally dealing with a few issues in the garden. Some birds love nibbling on lettuce leaves or freshly sprouted seeds. To protect the plants, I cover them with either bird netting or a floating row cover and that resolves the problem. Other than the lettuce issue, which usually happens throughout the season, I find that once little seedlings have passed that initial tender and tasty stage, the birds won't bother them and I can remove the covers.

The quail in my garden, however, think pea leaves are delicious. Since it's difficult to cover a whole pea patch, I hang mylar flash tape and/or toy snakes from the trellis to scare them away. Robins have a penchant for stealing cherries and berries, but flash tape and bird netting work well to keep them from landing in the trees or in the berry patch . . . most of the time.

The benefits and pleasure to be gained from bringing wild birds to your garden outweigh the hassle of protecting a few choice morsels from them. The bottom line for attracting our feathered friends is the same as for beneficials: If you provide them with a diverse landscape of trees, shrubs, grasses, perennials and annuals, they will come.

Friend or foe? Perhaps both! Find out on page 84.

2

MEET THE BUGS

Let's talk about bugs in the vegetable garden. It is always disheartening to discover damage from pests that covet your lovingly grown vegetables just as much as you do. The challenge is determining who the culprits are and knowing what to do about them.

This requires a bit of sleuthing on your part by observing the bugs—if you can see them—and the type of damage they are causing. Some are nocturnal and retreat to their hiding places during the day. In those cases, a trip to the garden after dark, with flashlight in hand, is in order. But more than anything, you'll want to be very observant and gather the facts.

To help you with this, I have assembled the Vegetable Crops and Potential Pest Problems table, which begins on page 36. Here's how it works.

Find the vegetable you are growing, then read through the descriptions of potential pest damage until you find one similar to what you've observed. Perhaps you're seeing chewed leaves, a sticky residue, holes in the fruits, or wilting leaves. Next to each type of damage are possible culprits.

The table also contains the plant family name for each crop. Why is this information important? Some pests are unique to, or regularly damage, the crops within specific plant families. If you are growing a vegetable that isn't included in the table but you know the plant family it belongs to, refer to another member of that family and read which pests target them. You'll find the plant family table and their members on page 23.

Once you have a list of possible suspects, go to each profile (the table provides the relevant page number). You'll see what it looks like, find a description, learn more about its life cycle, a list of crops it most typically attacks, signs of its activity, how to control it, and its natural predators.

I often find the description of a pest's life cycle helps me formulate a plan for outsmarting it. It can be as simple as adjusting the timing of a planting or using a barrier to prevent the adult pests from laying eggs on their favorite plants. If I can disrupt the life cycle of a particular pest—either by not planting its host crop for a couple of years or by targeting one of its growth stages—I will prevent future problems with it.

For example, cucumber beetles lay eggs on the soil beneath their host plants. If I place a thick mulch around the plants, the beetles won't be able to lay their eggs, which in turn means the next generation won't exist. A little knowledge goes a long way.

Let's get back to the Vegetable Crops and Potential Pest Problems table. I have a favor to ask of you: When you first look at it, please promise me you won't freak out! While there are a lot of pests that have the potential to damage your vegetable crops, that doesn't necessarily mean you're going to see every one of them in your garden.

Some pests are only present in certain regions or climates. In my Spokane, Washington, garden, our winters are too cold for Japanese beetles and Mexican bean beetles. While our short growing season can be challenging at times, I'm grateful those damaging beetles haven't come this far north . . . yet.

Other pests need the right conditions to be present in order to take up residence. This brings up a very important issue: As you read through the pest profiles, you'll notice that many of them have become highly resistant to pesticides. Some of the most damaging pests—such as Colorado potato beetles, diamondback caterpillars, and spider mites—are more prevalent in landscapes where synthetic chemicals are used, and misused, on a regular basis. While many insecticides are labeled to control specific pests, they kill their predators as well. Eliminating pesticide use is the number one way to encourage natural predators.

Here's the good news: All of the controls listed within each pest profile are organic. I've included cultural controls that greatly reduce or eliminate the chance for pests to be present in your garden. There are also simple techniques you can use to make it more difficult, if not impossible, for damaging pests to gain access to your crops. Every control is research-based, which means you can feel confident about using them. You'll find a description of each one beginning on page 136.

While you'll frequently see the recommendation to clean up plant debris at the end of the growing season, I also suggest you leave debris in place. Why the contradiction? Some beneficials—ladybugs, damsel bugs, and rove beetles, for example—overwinter in garden debris or leaf litter beneath trees and shrubs. Many solitary bees lay their eggs in the hollow stems of dead plants. As an organic gardener looking to create a balanced environment, you *want* to attract as many beneficials as possible.

Evaluate your garden: Does plant debris have a negligible impact or is it making troublesome pests feel right at home? The key here is location. The information included in the pest profiles will help you determine ways to accommodate the beneficials by providing habitat—in an area *away* from targeted crops.

As I mentioned in chapter one, we might not recognize some as allies. We squash them, thinking they were going to eat our crops. The profiles of beneficials beginning on page 124 will help you recognize the most common ones, find out which pests they eat, and discover how to encourage them in your landscape.

It is my hope that the more you learn about beneficials, the more you will look upon each trip out to the garden as an adventure. It is an opportunity to observe which ones are active depending upon the time of day or the season—and, if you're lucky, you'll see some in action.

In my own garden, I've made the happy discovery that far more beneficials have taken up residence than I thought possible. One morning, I found the coolest-looking bug crawling up a row cover on one of my raised beds. It turned out to be a snakefly (see page 130). Its distinctive appearance made it easy to research with the phrase "insect with long neck" and there it was at the top of the results! Snakeflies are beneficial because as adults they prey on aphids and other insects, and their larvae eat grubs. It would have been a terrible mistake if I had killed this new insect, thinking it was trying to get under the row cover to damage my broccoli.

Spiders tend to evoke fear in a lot of people, who then squish these incredibly helpful arachnids. While most spiders have venom, very few are harmful to humans. The vast majority of the spiders in our gardens are beneficial creatures that prey on pests such as aphids, cutworms, harlequin bugs, and leafminers. So please, leave spiders alone in the garden and let them do their important work. No matter which type of creature you encounter, get a good look at it, research it if need be, and make an informed decision.

I mentioned how a hand lens is a great identification tool, but don't overlook how useful smartphones can be for the same purpose. We carry one at all times, right? Many bugs are so small, it can be difficult to see their distinguishing features. Why not snap a photo with your phone and pop it into one of the many apps that can provide quick results to at least narrow it down? You can also zoom in on the picture for more detail. Even if you have trouble coming up with an ID right away, this gives you an opportunity to see how cool they are!

Are you ready to solve the "whodunit" case in your garden? Get started with the Vegetable Crops and Potential Pest Problems table, read the profile pages to determine the guilty party, and put together a plan of action.

TERMS TO KNOW

Before you dive into the pest profiles, here are definitions of some terms you'll be seeing.

Frass: Gardeners can often identify an insect based on the color and appearance of this excrement generated by the larval stage of an insect.

Honeydew: This sticky, sugary secretion from aphids and other insects often attracts other insects, especially ants. You won't find it nearly as attractive on your plants.

Instar: As they grow, insects, arachnids, and other types of bugs shed their exoskeletons and reach a new instar, which is a fancy name for a developmental stage on the way to sexual maturity. Some go through several instars.

Larva: The first stage of an insect after it has hatched from an egg; the plural form is larvae.

Nymph: An immature form of an insect goes through a simple metamorphosis; the nymph looks much like the adult version but is generally smaller and often lacks wings. Dragonflies, grasshoppers, and damselflies go through a nymph stage.

Prolegs: Sometimes called false legs, these are the fleshy limbs found on the abdomen of some caterpillars and other larvae.

Pupa: This life stage of certain types of insects prior to reaching their mature form often occurs within a chrysalis or capsule, and frequently within the soil, before the insect emerges as an adult. Insects that go through a pupal stage include butterflies, moths, and beetles. The plural form is pupae.

VEGETABLE CROPS & POTENTIAL PEST PROBLEMS

PLANT NAME	PLANT FAMILY	PROBLEMS/DAMAGE	POSSIBLE CULPRITS
Artichoke	Sunflower (*Asteraceae*)	Clusters of small insects on the undersides of leaves; honeydew; curled or yellow leaves; stunted growth; deformed buds	Aphid (page 60)
		Abnormal growth of leaves and flower stalks	Beet armyworm (page 64)
		Seedlings or plant stems cut off at the soil surface	Cutworm (page 80)
		Irregular holes in leaves	Earwig (page 84) Slug and snail (page 108)
		Tiny holes in leaves and stems; stunted growth	Flea beetle (page 86)
		Squiggly lines, blotchy areas, or clear patches in leaves	Leafminer (page 98)
		Slime trails	Slug and snail (page 108)
		Chewing damage to leaves and buds, accompanied by fine webbing	Spider mite (page 110)
		Holes in artichoke heads, damaged scales that turn brown, excrement	Stink bug (page 116)
		Mottled, stunted leaves; cloud of small white flies when the plant is disturbed; honeydew	Whitefly (page 120)
Arugula	Brassica (*Brassicaceae*)	Clusters of small insects on the undersides of leaves; honeydew; puckered, curling, or discolored leaves; stunted growth	Aphid (page 60)
		Skeletonized leaves	Beet armyworm (page 64) Mexican bean beetle (page 102)
		Feeding damage to the undersides of leaves; irregular holes in leaves; green excrement below the damage; green caterpillars on leaf midribs	Cabbage looper (page 68) Cabbage worm (page 70)

PLANT NAME	PLANT FAMILY	PROBLEMS/DAMAGE	POSSIBLE CULPRITS
Arugula	Brassica (*Brassicaceae*)	Seedlings or plant stems cut off at the soil surface	Cutworm (page 80)
		Holes in the leaves and stems; stunted seedlings	Diamondback caterpillar (page 82) Flea beetle (page 86)
		White blotches on leaves; wilting leaves or plants; brown foliage	Harlequin bug (page 90)
		Discoloration of plant stems, leaves, and flowers; distorted plant growth	Lygus bug (page 100)
		Irregular holes in leaves; slime trails	Slug and snail (page 108)
		Wilted, stunted, yellow leaves	Root maggot (page 106) Whitefly (page 120)
		Mottled leaves; cloud of small white flies when the plant is disturbed; honeydew	Whitefly (page 120)
Asparagus	Asparagus (*Asparagaceae*)	Clusters of small insects on the ferns; honeydew; stunted growth	Aphid (page 60)
		Dark eggs on the ferns; small beetles or greenish-gray larvae on the ferns; distorted, brown, or dead spears	Asparagus beetle (page 62)
		Pale green or pink eggs or clusters of eggs covered with white cottony scales on the ferns; green larvae with dark stripes feeding on the ferns	Beet armyworm (page 64)
		Chewed asparagus ferns	Cucumber beetle (page 78)
		Emerging spears have chew marks near the base	Cutworm (page 80)
		White spots in fern foliage; damaged spears	Harlequin bug (page 90)

PLANT NAME	PLANT FAMILY	PROBLEMS/DAMAGE	POSSIBLE CULPRITS
Bean	Legume (*Fabaceae*)	Clusters of small insects on the undersides of leaves; honeydew; puckered, curling, or discolored leaves; stunted growth	Aphid (page 60)
		Skeletonized leaves	Beet armyworm (page 64)
			Japanese beetle (page 94)
			Mexican bean beetle (page 102)
		Chewed flowers and/or leaves	Blister beetle (page 66)
		Holes bored into bean pods	Corn earworm (page 76)
			Mexican bean beetle (page 102)
			Slug and snail (page 108)
		Irregular holes in leaves and flowers; chewed seedlings; wilting plants	Cucumber beetle (page 78)
			Mexican bean beetle (page 102)
			Slug and snail (page 108)
		Seedlings or plant stems cut off at the soil surface	Cutworm (page 80)
		Irregular holes in leaves	Earwig (page 84)
			Grasshopper (page 88)
			Slug and snail (page 108)
		White blotches on leaves; wilting leaves or plants; brown foliage	Harlequin bug (page 90)
		Yellowed leaves; deformed growth	Leafhopper (page 96)
			Whitefly (page 120)
		Stunted growth	Leafhopper (page 96)
			Spider mite (page 110)
			Thrips (page 118)
			Wireworm (page 122)
		Squiggly lines, blotchy areas, or clear patches in leaves	Leafminer (page 98)

PLANT NAME	PLANT FAMILY	PROBLEMS/DAMAGE	POSSIBLE CULPRITS
Bean	Legume (*Fabaceae*)	Discoloration of plant stems, leaves, and flowers; distorted plant growth; pitted bean pods	Lygus bug (page 100)
		Chewed stems and leaves of seedlings	Pillbug and sowbug (page 104)
		Slime trails	Slug and snail (page 108)
		White or yellow stippling of leaves; fine webbing	Spider mite (page 110)
		Sunken surface of the bean seeds and pods	Stink bug (page 116)
		Stippled, silvery, or rolled leaves; black excrement	Thrips (page 118)
		Mottled leaves; leaves fall off; cloud of small white flies when plants are disturbed; honeydew	Whitefly (page 120)
		Wilting plants	Wireworm (page 122)
Beet	Beet (*Amaranthaceae*)	Clusters of small insects on the undersides of leaves; honeydew; puckered, curling, or discolored leaves; stunted growth	Aphid (page 60)
		Skeletonized leaves	Beet armyworm (page 64)
		Irregular holes in leaves	Cucumber beetle (page 78)
			Slug and snail (page 108)
		Small holes in leaves and stems	Flea beetle (page 86)
		Yellowed leaves; stunted, deformed growth	Leafhopper (page 96)
		Squiggly lines, blotchy areas, or clear patches in leaves	Leafminer (page 98)
		Discoloration of plant stems, leaves, and flowers; distorted plant growth	Lygus bug (page 100)
		Scarred surface of roots; tunnels through the roots	Root maggot (page 106)
		Slime trails	Slug and snail (page 108)

PLANT NAME	PLANT FAMILY	PROBLEMS/DAMAGE	POSSIBLE CULPRITS
Broccoli	Brassica (*Brassicaceae*)	Clusters of small insects on the undersides of leaves; honeydew; puckered, curling, or discolored leaves; stunted growth	Aphid (page 60)
		Tunnels bored into the broccoli heads	Beet armyworm (page 64) Cabbage looper (page 68)
		Feeding damage to the undersides of leaves; irregular holes in leaves; green excrement below the damage; green caterpillars on leaf midribs	Cabbage looper (page 68) Cabbage worm (page 70) Diamondback caterpillar (page 82)
		Seedlings or plant stems cut off at the soil surface	Cutworm (page 80)
		Small holes in leaves and stems	Flea beetle (page 86)
		White blotches on leaves; wilting leaves or plants; brown foliage	Harlequin bug (page 90)
		Discoloration of plant stems, leaves, and flowers; distorted plant growth	Lygus bug (page 100)
		Wilted, stunted, yellow leaves	Root maggot (page 106) Whitefly (page 120)
		Irregular holes in leaves; slime trail	Slug and snail (page 108)
		Cloud of small white flies when the plant is disturbed; honeydew	Whitefly (page 120)
Brussels sprouts	Brassica (*Brassicaceae*)	Clusters of small insects on the undersides of leaves; honeydew; puckered, curling, or discolored leaves; stunted growth	Aphid (page 60)
		Tunnels bored into the Brussels sprouts heads	Beet armyworm (page 64) Cabbage looper (page 68)
		Feeding damage to the undersides of leaves; irregular holes in leaves; green excrement below the damage; green caterpillars on leaf midribs	Cabbage looper (page 68) Cabbage worm (page 70)

PLANT NAME	PLANT FAMILY	PROBLEMS/DAMAGE	POSSIBLE CULPRITS
Brussels sprouts	Brassica (*Brassicaceae*)	Seedlings or plant stems cut off at the soil surface	Cutworm (page 80)
		Irregular holes in the leaves and stems; stunted seedlings; chewing damage to Brussels sprouts heads	Diamondback caterpillar (page 82)
		Small holes in leaves and stems	Flea beetle (page 86)
		White blotches on leaves; wilting leaves or plants; brown foliage	Harlequin bug (page 90)
		Wilted, stunted, yellow leaves	Root maggot (page 106) Whitefly (page 120)
		Irregular holes in leaves; slime trail	Slug and snail (page 108)
		Mottled leaves; leaves fall off; cloud of small white flies when the plant is disturbed; honeydew	Whitefly (page 120)
Cabbage	Brassica (*Brassicaceae*)	Clusters of small insects on the undersides of leaves; honeydew; puckered, curling, or discolored leaves; stunted growth	Aphid (page 60)
		Tunnels bored into cabbage heads	Beet armyworm (page 64) Cabbage looper (page 68)
		Feeding damage to the undersides of leaves; irregular holes in leaves; green excrement below the damage; green caterpillars on leaf midribs; chewing damage to cabbage heads	Cabbage looper (page 68) Cabbage worm (page 70) Diamondback caterpillar (page 82)
		Seedlings or plant stems cut off at the soil surface	Cutworm (page 80)
		Small holes in leaves and stems	Flea beetle (page 86)
		White blotches on leaves; wilting leaves or plants; brown foliage	Harlequin bug (page 90)
		Discoloration of plant stems, leaves, and flowers; distorted plant growth	Lygus bug (page 100)
		Irregular holes in leaves; chewed seedlings; wilting plants; skeletonized leaves	Mexican bean beetle (page 102)
		Wilted, stunted, yellow leaves; plant dies suddenly	Root maggot (page 106)

PLANT NAME	PLANT FAMILY	PROBLEMS/DAMAGE	POSSIBLE CULPRITS
Cabbage	Brassica (*Brassicaceae*)	Irregular holes in leaves; slime trail	Slug and snail (page 108)
		Stunted growth; yellow or mottled leaves; leaves fall off; cloud of small white flies when the plant is disturbed; honeydew	Whitefly (page 120)
Carrot	Carrot (*Apiaceae*)	Root damage (tunnels, grooves, holes, scarring of the surface)	Carrot rust fly (page 72) Root maggot (page 106) Wireworm (page 122)
		Rust-colored excrement; wilting or dead plants	Carrot rust fly (page 72)
		Seedlings or plant stems cut off at the soil surface	Cutworm (page 80)
		Chewed foliage	Grasshopper (page 88)
		Discoloration of plant stems and leaves; distorted plant growth	Lygus bug (page 100)
Cauliflower	Brassica (*Brassicaceae*)	Clusters of small insects on the undersides of leaves; honeydew; curled or yellow leaves; stunted growth; deformed buds	Aphid (page 60)
		Skeletonized leaves; tunneling into the cauliflower heads	Beet armyworm (page 64)
		Feeding damage to the undersides of leaves; green excrement below the damage; green caterpillars on leaf midribs; tunnels bored into the cauliflower heads	Cabbage looper (page 68) Cabbage worm (page 70) Diamondback caterpillar (page 82)
		Irregular holes in leaves and/or stems	Cabbage looper (page 68) Cabbage worm (page 70) Diamondback caterpillar (page 82) Slug and snail (page 108)

PLANT NAME	PLANT FAMILY	PROBLEMS/DAMAGE	POSSIBLE CULPRITS
Cauliflower	Brassica (*Brassicaceae*)	Tiny holes in leaves and stems; stunted growth	Flea beetle (page 86)
		White blotches on leaves; wilting leaves or plants; brown foliage	Harlequin bug (page 90)
		Discoloration of plant stems, leaves, and flowers; distorted plant growth	Lygus bug (page 100)
		Wilted, stunted, yellow leaves	Root maggot (page 106) Whitefly (page 120)
		Slime trails	Slug and snail (page 108)
		Cloud of small white flies when the plant is disturbed; honeydew	Whitefly (page 120)
Corn	Grass (*Poaceae*)	Clusters of small insects on the undersides of leaves; honeydew; puckered, curling, or discolored leaves; stunted growth	Aphid (page 60)
		Irregular holes in leaves; damaged ears of corn	Beet armyworm (page 64) Earwig (page 84) Grasshopper (page 88) Stink bug (page 116)
		Damage to corn ears, silks, and/or tassels	Beet armyworm (page 64) Corn earworm (page 76) Cucumber beetle (page 78) Earwig (page 84) Grasshopper (page 88) Japanese beetle (page 94) Stink bug (page 116)
		Seedlings or plant stems cut off at the soil surface	Cutworm (page 80)
		Tiny holes in leaves and stems; stunted growth	Flea beetle (page 86)

PLANT NAME	PLANT FAMILY	PROBLEMS/DAMAGE	POSSIBLE CULPRITS
Corn	Grass (*Poaceae*)	Skeletonized leaves	Japanese beetle (page 94)
		Stunted or wilting plants	Wireworm (page 122)
Cucumber	Cucurbit (*Cucurbitaceae*)	Irregular holes in leaves and flowers; chewed seedlings; wilting plants; chewing damage on skins of fruits	Cucumber beetle (page 78)
		White blotches on leaves; wilting leaves or plants; brown foliage	Harlequin bug (page 90)
		Discoloration of plant stems, leaves, and flowers; distorted plant growth	Lygus bug (page 100)
		Chewed stems and leaves of seedlings	Pillbug and sowbug (page 104)
		White or yellow stippling of the leaves; dying plants; fine webbing	Spider mite (page 110)
		Stunted growth	Spider mite (page 110) Thrips (page 118) Whitefly (page 120) Wireworm (page 122)
		Tiny specks on leaves that turn yellow and brown; wilting leaves; plant dies; reddish-brown eggs on stems or leaf undersides; nymphs and adults that scatter quickly	Squash bug (page 112)
		Stippled, silvery, rolled leaves; black excrement	Thrips (page 118)
		Cloud of small white flies when the plant is disturbed; honeydew	Whitefly (page 120)
		Wilting plants	Wireworm (page 122)

PLANT NAME	PLANT FAMILY	PROBLEMS/DAMAGE	POSSIBLE CULPRITS
Eggplant	Nightshade (*Solanaceae*)	Feeding damage on leaves, flowers, and/or fruits	Beet armyworm (page 64) Blister beetle (page 66) Colorado potato beetle (page 74) Flea beetle (page 86) Hornworm (page 92) Japanese beetle (page 94)
		Dark excrement; yellow or orange eggs on the undersides of leaves	Colorado potato beetle (page 74)
		Tiny holes in leaves and stems; stunted growth; wilting plants	Flea beetle (page 86)
		White blotches on leaves; wilting leaves or plants; brown foliage	Harlequin bug (page 90)
		Skeletonized leaves	Japanese beetle (page 94)
		Yellow leaves; stunted, deformed growth	Leafhopper (page 96) Whitefly (page 120)
		Discoloration of plant stems, leaves, and flowers; distorted plant growth; flowers and fruits drop prematurely; "cat-facing" damage (scarring and cavities)	Lygus bug (page 100)
		Defoliated plants; dark excrement on leaves and below feeding damage	Hornworm (page 92)
		Cloud of small white flies when the plant is disturbed; honeydew	Whitefly (page 120)
Kale	Brassica (*Brassicaceae*)	Clusters of small insects on the undersides of leaves; honeydew; curled or yellow leaves; stunted growth; deformed buds	Aphid (page 60)
		Skeletonized leaves	Beet armyworm (page 64) Mexican bean beetle (page 102)

PLANT NAME	PLANT FAMILY	PROBLEMS/DAMAGE	POSSIBLE CULPRITS
Kale	Brassica (*Brassicaceae*)	Feeding damage to the undersides of leaves; irregular holes in leaves; green excrement below the damage; green caterpillars on leaf midribs	Cabbage looper (page 68) Cabbage worm (page 70)
		Seedlings or plant stems cut off at the soil surface	Cutworm (page 80)
		Irregular holes in leaves and stems; stunted seedlings	Diamondback caterpillar (page 82) Mexican bean beetle (page 102)
		Tiny holes in leaves and stems; stunted growth; wilting plants	Flea beetle (page 86)
		White blotches on leaves; wilting leaves or plants; brown foliage	Harlequin bug (page 90)
		Discoloration of plant stems, leaves, and flowers; distorted plant growth	Lygus bug (page 100)
		Chewed seedlings; wilting plants	Mexican bean beetle (page 102)
		Wilted, stunted, yellow leaves; plant dies suddenly	Root maggot (page 106)
Kohlrabi	Brassica (*Brassicaceae*)	Clusters of small insects on the undersides of leaves; honeydew; curled or yellow leaves; stunted growth; deformed buds	Aphid (page 60)
		Skeletonized leaves; tunneling into kohlrabi heads	Beet armyworm (page 64)
		Feeding damage to the undersides of leaves; irregular holes in leaves; green excrement below the damage; green caterpillars on leaf midribs; stunted seedlings	Cabbage looper (page 68) Cabbage worm (page 70) Diamondback caterpillar (page 82)
		Seedlings or plant stems cut off at the soil surface	Cutworm (page 80)
		Tiny holes in leaves and stems; stunted growth; wilting plants	Flea beetle (page 86)

PLANT NAME	PLANT FAMILY	PROBLEMS/DAMAGE	POSSIBLE CULPRITS
Kohlrabi	Brassica (*Brassicaceae*)	White blotches on leaves; wilting leaves or plants; brown foliage	Harlequin bug (page 90)
		Discoloration of plant stems, leaves, and flowers; distorted plant growth	Lygus bug (page 100)
		Wilted, stunted, yellow leaves; plant dies suddenly	Root maggot (page 106)
Lettuce	Sunflower (*Asteraceae*)	Clusters of small insects on the undersides of leaves; honeydew; curled or yellow leaves; stunted growth	Aphid (page 60)
		Skeletonized leaves	Beet armyworm (page 64) / Japanese beetle (page 94)
		Feeding damage to the undersides of leaves; irregular holes in leaves; green excrement below the damage; green caterpillars on leaf midribs	Cabbage looper (page 68)
		Tunnels bored into the lettuce heads	Corn earworm (page 76)
		Seedlings or plant stems cut off at the soil surface	Cutworm (page 80)
		Irregular holes in leaves	Earwig (page 84) / Grasshopper (page 88) / Slug and snail (page 108)
		Tiny holes in leaves and stems	Flea beetle (page 86)
		Stunted or wilting plants	Spider mite (page 110) / Wireworm (page 122)
		Chewed stems and leaves of seedlings	Pillbug and sowbug (page 104)
		Slime trails	Slug and snail (page 108)
		White or yellow stippling of the leaves; fine webbing	Spider mite (page 110)

PLANT NAME	PLANT FAMILY	PROBLEMS/DAMAGE	POSSIBLE CULPRITS
Melon (including watermelon)	Cucurbit (*Cucurbitaceae*)	Irregular holes in leaves and flowers; chewed seedlings; chewing damage on skins of fruits	Cucumber beetle (page 78)
		Stunted, distorted, or wilting plants	Cucumber beetle (page 78) Harlequin bug (page 90) Lygus bug (page 100) Spider mite (page 110) Squash bug (page 112) Thrips (page 118) Wireworm (page 122)
		White blotches on leaves; brown foliage	Harlequin bug (page 90)
		Squiggly lines, blotchy areas, or clear patches in leaves	Leafminer (page 98)
		Discoloration of plant stems, leaves, and flowers	Lygus bug (page 100)
		Chewed stems and leaves of seedlings	Pillbug and sowbug (page 104)
		White or yellow stippling of the leaves; dying plants; fine webbing	Spider mite (page 110)
		Tiny specks on leaves that turn yellow and brown; plant dies; reddish-brown eggs on stems or leaf undersides; nymphs and adults that scatter quickly	Squash bug (page 112)
		Stippled, silvery, rolled leaves; black excrement	Thrips (page 118)

PLANT NAME	PLANT FAMILY	PROBLEMS/DAMAGE	POSSIBLE CULPRITS
Mustard	Brassica (*Brassicaceae*)	Clusters of small insects on the undersides of leaves; honeydew; curled or yellow leaves; stunted growth	Aphid (page 60)
		Skeletonized leaves	Beet armyworm (page 64)
			Japanese beetle (page 94)
		Feeding damage to the undersides of leaves; irregular holes in leaves; green excrement below the damage; green caterpillars on leaf midribs	Cabbage looper (page 68)
			Cabbage worm (page 70)
			Diamondback caterpillar (page 82)
		Irregular holes in the leaves and stems	Slug and snail (page 108)
		Stunted or wilting plants	Flea beetle (page 86)
			Harlequin bug (page 90)
			Lygus bug (page 100)
			Mexican bean beetle (page 102)
			Root maggot (page 106)
			Whitefly (page 120)
		Tiny holes in leaves and stems	Flea beetle (page 86)
		White blotches on leaves; brown foliage	Harlequin bug (page 90)
		Discoloration of plant stems, leaves, and flowers	Lygus bug (page 100)
		Chewed seedlings	Mexican bean beetle (page 102)
			Pillbug and sowbug (page 104)
		Yellow leaves; plant dies suddenly	Root maggot (page 106)
		Slime trails	Slug and snail (page 108)
		Cloud of small white flies when the plant is disturbed; honeydew	Whitefly (page 120)

PLANT NAME	PLANT FAMILY	PROBLEMS/DAMAGE	POSSIBLE CULPRITS
Onion	Onion (*Amaryllidaceae*)	Clusters of small insects on leaves; honeydew	Aphid (page 60)
		Feeding damage on leaves	Beet armyworm (page 64)
			Grasshopper (page 88)
			Pillbug and sowbug (page 104)
		White blotches on leaves; wilting leaves or plants; brown foliage	Harlequin bug (page 90)
		Scarred root surfaces; tunnels through the roots	Root maggot (page 106)
		Silver streaks or white blotches on leaves; brown leaf tips	Thrips (page 118)
		Stunted or wilting plants	Wireworm (page 122)
Pea	Legume (*Fabaceae*)	Clusters of small insects on the undersides of leaves; honeydew; curled or yellow leaves; stunted growth; deformed buds	Aphid (page 60)
		Feeding damage on leaves, stems, and/or flowers	Beet armyworm (page 64)
			Blister beetle (page 66)
			Cabbage looper (page 68)
			Pillbug and sowbug (page 104)
		Feeding damage to the undersides of leaves; irregular holes in leaves; green excrement below the damage; green caterpillars on leaf midribs	Cabbage looper (page 68)
		Seedlings or plant stems cut off at the soil surface	Cutworm (page 80)
		White blotches on leaves; wilting leaves or plants; brown foliage	Harlequin bug (page 90)
		Squiggly lines, blotchy areas, or clear patches in leaves	Leafminer (page 98)

PLANT NAME	PLANT FAMILY	PROBLEMS/DAMAGE	POSSIBLE CULPRITS
Pea	Legume (*Fabaceae*)	Discoloration of plant stems, leaves, and flowers; distorted plant growth	Lygus bug (page 100)
		Stunted or wilting plants	Spider mite (page 110) Thrips (page 118) Wireworm (page 122)
		White or yellow stippling of the leaves; fine webbing	Spider mite (page 110)
		Stippled, silvery, or rolled leaves; black excrement	Thrips (page 118)
Pepper	Nightshade (*Solanaceae*)	Feeding damage on leaves and/or flowers	Beet armyworm (page 64) Blister beetle (page 66) Colorado potato beetle (page 74) Earwig (page 84) Flea beetle (page 86) Hornworm (page 92) Japanese beetle (page 94)
		Feeding damage to fruits	Beet armyworm (page 64) Corn earworm (page 76) Earwig (page 84) Hornworm (page 92)
		Excrement on plants	Beet armyworm (page 64) Colorado potato beetle (page 74) Hornworm (page 92)

PLANT NAME	PLANT FAMILY	PROBLEMS/DAMAGE	POSSIBLE CULPRITS
Pepper	Nightshade (*Solanaceae*)	Pale green eggs on the undersides of leaves (hornworm); yellow or orange eggs on the undersides of leaves (beetle)	Colorado potato beetle (page 74)
			Hornworm (page 92)
		Seedlings or plant stems cut off at the soil surface	Cutworm (page 80)
		Stunted or deformed growth	Flea beetle (page 86)
			Leafhopper (page 96)
			Lygus bug (page 100)
		White blotches on leaves; wilting leaves or plants; brown foliage	Harlequin bug (page 90)
		Skeletonized leaves	Japanese beetle (page 94)
		Yellow leaves	Leafhopper (page 96)
		Squiggly lines, blotchy areas, or clear patches in leaves	Leafminer (page 98)
		Discoloration of plant stems, leaves, and flowers; flowers and fruits drop prematurely	Lygus bug (page 100)
Potato	Nightshade (*Solanaceae*)	Skeletonized leaves	Beet armyworm (page 64)
		Chewed flowers and/or leaves	Blister beetle (page 66)
		Feeding damage to the undersides of leaves; irregular holes in leaves; green frass below the damage; green caterpillars on leaf midribs	Cabbage looper (page 68)
		Chewed foliage; dark excrement; yellow or orange eggs on the undersides of leaves	Colorado potato beetle (page 74)
		Irregular holes in leaves and flowers; wilting plants	Cucumber beetle (page 78)
		Irregular holes in leaves	Earwig (page 84)
		Tiny holes in leaves and stems; stunted growth; scars on potato tubers	Flea beetle (page 86)
		White blotches on leaves; wilting leaves or plants; brown foliage	Harlequin bug (page 90)
		Yellowed leaves; stunted, deformed growth	Leafhopper (page 96)

PLANT NAME	PLANT FAMILY	PROBLEMS/DAMAGE	POSSIBLE CULPRITS
Potato	Nightshade (*Solanaceae*)	Squiggly lines, blotchy areas, or clear patches in leaves	Leafminer (page 98)
		Discoloration of plant stems, leaves, and flowers; distorted plant growth	Lygus bug (page 100)
		Chewed foliage and fruits; completely defoliated plants; dark excrement on leaves and below feeding damage	Hornworm (page 92)
		Yellow or stunted leaves; cloud of small white flies when the plant is disturbed; honeydew	Whitefly (page 120)
		Straight, round holes in potato tubers	Wireworm (page 122)
Pumpkin	Cucurbit (*Cucurbitaceae*)	Irregular holes in leaves and flowers; chewed seedlings; wilting plants; chewing damage on skins of fruits	Cucumber beetle (page 78)
		White blotches on leaves; wilting leaves or plants; brown foliage	Harlequin bug (page 90)
		Discoloration of plant stems, leaves, and flowers; distorted plant growth	Lygus bug (page 100)
		Tiny specks on leaves that turn yellow and brown; wilting leaves; plant dies; reddish-brown eggs on stems or leaf undersides; nymphs and adults that scatter quickly	Squash bug (page 112)
		Yellowing leaves; wilting plants; puncture holes in stems near the base of the plant; golden excrement; dead vines	Squash vine borer (page 114)
		Stunted growth; stippled, silvery, or rolled leaves; black excrement	Thrips (page 118)
Radish	Brassica (*Brassicaceae*)	Clusters of small insects on the undersides of leaves; honeydew; curled or yellow leaves; stunted growth	Aphid (page 60)
		Skeletonized leaves	Beet armyworm (page 64)
		Feeding damage to the undersides of leaves; irregular holes in leaves; green excrement below the damage; green caterpillars on leaf midribs	Cabbage looper (page 68)
			Cabbage worm (page 70)
			Diamondback caterpillar (page 82)

PLANT NAME	PLANT FAMILY	PROBLEMS/DAMAGE	POSSIBLE CULPRITS
Radish	Brassica (*Brassicaceae*)	Seedlings or plant stems cut off at the soil surface	Cutworm (page 80)
		Tiny holes in leaves and stems; stunted growth; wilting plants	Flea beetle (page 86)
		White blotches on leaves; wilting leaves or plants; brown foliage	Harlequin bug (page 90)
		Discoloration of plant stems, leaves, and flowers; distorted plant growth	Lygus bug (page 100)
		Scarred surface of roots; tunnels through the roots	Root maggot (page 106)
		Irregular holes in leaves; slime trails	Slug and snail (page 108)
		Yellow or stunted leaves; cloud of small white flies when the plant is disturbed; honeydew	Whitefly (page 120)
Rutabaga	Brassica (*Brassicaceae*)	Clusters of small insects on the undersides of leaves; honeydew; curled or yellow leaves; stunted growth	Aphid (page 60)
		Skeletonized leaves	Beet armyworm (page 64)
		Feeding damage to the undersides of leaves; irregular holes in leaves; green excrement below the damage; green caterpillars on leaf midribs	Cabbage looper (page 68) / Cabbage worm (page 70) / Diamondback caterpillar (page 82)
		Seedlings or plant stems cut off at the soil surface	Cutworm (page 80)
		Tiny holes in leaves and stems; stunted growth; wilting plants	Flea beetle (page 86)
		White blotches on leaves; wilting leaves or plants; brown foliage	Harlequin bug (page 90)
		Discoloration of plant stems, leaves, and flowers; distorted plant growth	Lygus bug (page 100)
		Scarred root surfaces; tunnels through the roots	Root maggot (page 106)

PLANT NAME	PLANT FAMILY	PROBLEMS/DAMAGE	POSSIBLE CULPRITS
Rutabaga	Brassica (*Brassicaceae*)	Irregular holes in leaves; slime trails	Slug and snail (page 108)
		Yellow or stunted leaves; cloud of small white flies when the plant is disturbed; honeydew	Whitefly (page 120)
Spinach	Beet (*Amaranthaceae*)	Skeletonized leaves	Beet armyworm (page 64)
		Feeding damage to the undersides of leaves; irregular holes in leaves; green excrement below the damage; green caterpillars on leaf midribs	Cabbage looper (page 68)
		Tiny holes in leaves and stems; stunted growth; wilting plants	Flea beetle (page 86)
		Large, irregular holes in the leaves	Grasshopper (page 88) Slug and snail (page 108)
		Skeletonized leaves	Japanese beetle (page 94)
		Squiggly lines, blotchy areas, or clear patches in leaves	Leafminer (page 98)
		Discoloration of plant stems, leaves, and flowers; distorted plant growth	Lygus bug (page 100)
		Slime trails	Slug and snail (page 108)
Squash (summer and winter)	Cucurbit (*Cucurbitaceae*)	Irregular holes in leaves and flowers; chewed seedlings; wilting plants; chewing damage on skins of fruits	Cucumber beetle (page 78)
		White blotches on leaves; wilting leaves or plants; brown foliage	Harlequin bug (page 90)
		Squiggly lines, blotchy areas, or clear patches in leaves	Leafminer (page 98)
		Discoloration of plant stems, leaves, and flowers; distorted plant growth	Lygus bug (page 100)
		Chewed stems and leaves of seedlings	Pillbug and sowbug (page 104)
		Tiny specks on leaves that turn yellow and brown; wilting leaves; plant dies; reddish-brown eggs on stems or leaf undersides; nymphs and adults that scatter quickly	Squash bug (page 112)

PLANT NAME	PLANT FAMILY	PROBLEMS/DAMAGE	POSSIBLE CULPRITS
Squash (summer and winter)	Cucurbit (*Cucurbitaceae*)	Yellowing leaves; wilting plants; puncture holes in stems near the base of the plant; golden excrement; dead vines	Squash vine borer (page 114)
		Stunted growth; stippled, silvery, or rolled leaves; black excrement	Thrips (page 118)
		Yellow or stunted leaves; cloud of small white flies when the plant is disturbed; honeydew	Whitefly (page 120)
Swiss chard	Beet (*Amaranthaceae*)	Skeletonized leaves	Beet armyworm (page 64)
		Tiny holes in leaves and stems; stunted growth; wilting plants	Flea beetle (page 86)
		Squiggly lines, blotchy areas, or clear patches in leaves	Leafminer (page 98)
		Discoloration of plant stems, leaves, and flowers; distorted plant growth	Lygus bug (page 100)
		Irregular holes in leaves; slime trails	Slug and snail (page 108)
Tomato	Nightshade (*Solanaceae*)	Skeletonized leaves	Beet armyworm (page 64) Japanese beetle (page 94)
		Chewed leaves and/or flowers	Blister beetle (page 66) Colorado potato beetle (page 74) Cucumber beetle (page 78) Flea beetle (page 86) Hornworm (page 92) Japanese beetle (page 94) Lygus bug (page 100) Pillbug and sowbug (page 104) Slug and snail (page 108)

PLANT NAME	PLANT FAMILY	PROBLEMS/DAMAGE	POSSIBLE CULPRITS
Tomato	Nightshade (*Solanaceae*)	Seedlings or plant stems cut off at the soil surface	Cutworm (page 80)
		Fruit damage	Blister beetle (page 66) Corn earworm (page 76) Hornworm (page 92) Lygus bug (page 100) Stink bug (page 116) Whitefly (page 120)
		Pale green eggs on the undersides of leaves (hornworm); yellow or orange eggs on the undersides of leaves (beetle)	Colorado potato beetle (page 74) Hornworm (page 92)
		Dark excrement near or below the damage	Colorado potato beetle (page 74) Hornworm (page 92)
		Stunted growth; wilting plants	Flea beetle (page 86) Lygus bug (page 100) Spider mite (page 110)
		White blotches on leaves; wilting leaves or plants; brown foliage	Harlequin bug (page 90)
		Yellowed leaves	Leafhopper (page 96)
		Squiggly lines, blotchy areas, or clear patches in leaves	Leafminer (page 98)
		Discoloration of plant stems, leaves, and flowers	Lygus bug (page 100)
		Slime trails	Slug and snail (page 108)
		White or yellow stippling of the leaves; fine webbing	Spider mite (page 110)
		Cloud of small white flies when the plant is disturbed; honeydew; uneven ripening of fruits	Whitefly (page 120)

PLANT NAME	PLANT FAMILY	PROBLEMS/DAMAGE	POSSIBLE CULPRITS
Turnip	Brassica (*Brassicaceae*)	Clusters of small insects on the undersides of leaves; honeydew; curled or yellow leaves	Aphid (page 60)
		Stunted or distorted growth	Aphid (page 60) Diamondback caterpillar (page 82) Flea beetle (page 86) Lygus bug (page 100) Whitefly (page 120)
		Skeletonized leaves	Beet armyworm (page 64)
		Feeding damage to the undersides of leaves; irregular holes in leaves; green frass below the damage; green caterpillars on leaf midribs	Cabbage looper (page 68) Cabbage worm (page 70)
		Irregular holes in leaves and stems	Diamondback caterpillar (page 82) Slug and snail (page 108)
		Tiny holes in leaves and stems	Flea beetle (page 86)
		White blotches on leaves; wilting leaves or plants; brown foliage	Harlequin bug (page 90)
		Discoloration of plant stems, leaves, and flowers	Lygus bug (page 100)
		Scarred root surfaces; tunnels through the roots	Root maggot (page 106)
		Slime trails	Slug and snail (page 108)
		Yellow leaves; cloud of small white flies when the plant is disturbed; honeydew	Whitefly (page 120)

Row covers are useful tools for deterring many different pests. Learn more on page 145.

APHID

Aphids grow up to ⅛ inch (3 mm) long and have soft, pear-shaped bodies that can be green, gray, white, yellow, brown, red, or black. Usually wingless, they have piercing/sucking mouthparts to drink sap from plant leaves, stems, and roots. The two cornicles at the end of an aphid's abdomen resemble horns. Aphids exude a waxy substance through these cornicles as a defense mechanism to repel predators that disturb them. They also excrete honeydew, a sugary liquid that attracts ants, which harvest the substance and protect the aphids from predators. While all of the many species of aphids adversely impact vegetable crops, the good news is that the controls for all of them are the same.

⅛" (3 mm)

Family name: Aphididae
Latin name: Many different species

You will never see a lone aphid on a plant!

LIFE CYCLE

Besides mating and laying eggs late in the growing season, aphids reproduce asexually. Some species hatch as winged young—primarily females—from fertilized eggs laid on nearby perennial plants the previous fall, then fly to the nearest host vegetable plant. Each female is essentially pregnant when she hatches and soon gives live birth to up to 60 more offspring, which have no wings. Root aphids have a similar life cycle and access plant roots through cracks in the soil. There are also aphid species that only live on their vegetable host plants.

TYPICALLY SEEN ON

Artichokes, asparagus, beans, beets, cabbage family crops (arugula, broccoli, Brussels sprouts, cabbage, cauliflower, kale, kohlrabi, mustard, radishes, rutabagas, turnips), corn, lettuce, onions, peas

Aphids excrete honeydew, which is a sticky, sugary liquid.

Aphids caused the puckering and discoloration on this kohlrabi leaf.

SIGNS OF THEIR ACTIVITY

Puckered, curling leaves; discoloration of leaves; stunted growth tips; honeydew; ants on or below plants; tiny insects, usually on the undersides of the leaves

NATURAL PREDATORS

Assassin bugs, big-eyed bugs, damsel bugs, earwigs, ground beetles, hoverflies, lacewings, ladybugs, minute pirate bugs, parasitic wasps, praying mantids, robber flies, soldier beetles, spiders, syrphid flies

CONTROLS

> Prior to planting, consider covering the soil with a reflective silver plastic mulch, which will make it more difficult for aphids to find host plants (refer to DIY on page 178).

> Crush the aphids (you may want to wear gloves).

> Blast them off of the plants with a jet of water.

> Use floating row covers over susceptible crops that do not require pollination.

> Let beneficials handle the problem for you.

> Avoid excessive use of high-nitrogen fertilizers, which produce the tender new growth that attract aphids.

> Apply diatomaceous earth, horticultural oil, insecticidal soaps, Neem, plant extracts, or pyrethrins.

> Do not purchase and release ladybugs, because they won't remain on your property and are likely non-natives that will adversely impact your local, native ladybugs.

MEET THE BUGS

ASPARAGUS BEETLE, COMMON AND SPOTTED

The common asparagus beetle and its larvae can cause a lot of damage to asparagus ferns and developing spears. The ¼-inch-long (6 mm) adults have red bodies with red-rimmed black wings punctuated by six white spots. The greenish-gray larvae have black heads and are ¹⁄₁₆ to ⅛ inch (2 to 3 mm) long. Larvae chew on the ferns, which interferes with the plant's ability to conduct photosynthesis and can kill the plants.

The spotted asparagus beetle and its larvae are much less damaging. The adults are orange with twelve black spots. The larvae, which are also orange, feed on the asparagus berries and do not damage the plant's spears or foliage. Gardeners sometimes mistake spotted asparagus beetles for adult ladybugs. Just remember that ladybugs have short antennae, less noticeable heads, and very round bodies compared to the long and narrow bodies of the asparagus beetles.

¼" (6 mm)

Family name: Chrysomelidae
Latin name: *Crioceris asparagi* (common), *Crioceris duodecimpunctata* (spotted)

Spotted asparagus beetle

LIFE CYCLE

Common asparagus beetles overwinter in debris near asparagus beds and mate in the spring when the asparagus spears are emerging. They lay eggs on the ferns, developing spears, or flowers. The eggs hatch in about a week and the larvae begin feeding. After two weeks, they fall to the ground to pupate, then emerge as adults one week later. Spotted asparagus beetles emerge a bit later in the spring and lay eggs on the foliage or asparagus berries. After hatching, the larvae feed on the berries.

TYPICALLY SEEN ON

Asparagus plants

Common asparagus beetle

Common asparagus beetle eggs

SIGNS OF THEIR ACTIVITY

Dark eggs laid on end on the ferns and/or green larvae (common asparagus beetle); orange larvae (spotted asparagus beetle); either type of beetle on the ferns or spears; distorted, brown, or completely dead spears

NATURAL PREDATORS

Lacewing larvae, ladybug larvae, parasitic wasp larvae, praying mantids, spiders

CONTROLS

> As asparagus spears emerge, look for eggs, larvae, and beetles; handpick them.

> Note that asparagus beetles are most active in the afternoon.

> Clean up garden debris to eliminate hiding places for beetles.

> Place a floating row cover over spears.

> Harvest young spears frequently.

> If spotted asparagus beetles are a problem in your region, consider planting male hybrids of asparagus, which don't produce berries, or picking the berries right away to eliminate the food source for the larvae.

> Apply beneficial nematodes, kaolin clay, Neem, plant extracts, pyrethrins, or spinosad.

> Avoid using pesticides since they kill natural predators.

BEET ARMYWORM

Despite their name, beet armyworms are damaging to a wide variety of vegetable crops, not just beets. These insects are most prevalent in areas with mild winters, although they will occasionally migrate to colder regions. The adult moth has mottled gray and brown forewings, pale brown translucent hindwings, and a 1-inch (2.5 cm) wingspan. Their eggs are pale green or pink. The larvae have smooth green skin, dark and wavy light stripes along the length of their bodies, and sometimes develop dark spots on the thorax. They also have four pairs of prolegs. Mature larvae are 1½ inches (3.8 cm) long.

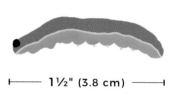

├─── 1½" (3.8 cm) ───┤

Family name: Noctuidae
Latin name:
Spodoptera exigua

Beet armyworm moth

LIFE CYCLE

Beet armyworm pupae overwinter in small capsules within the soil and emerge in the spring. After mating, each female lays a cluster of 80 to 150 eggs on the underside of a leaf and covers them with fuzzy white scales that look like cotton. While the moths only live about 10 days, they can lay up to 600 eggs in that short time. The larvae hatch in a couple of days and remain in groups while skeletonizing the leaves. As they mature, they also burrow into plants and easily move from one plant to another. Their complete life cycle takes place in 24 to 30 days and there are up to six generations per year, depending on the weather conditions.

TYPICALLY SEEN ON

Artichokes, asparagus, beans, beet family crops (beets, spinach, Swiss chard), cabbage family crops (arugula, broccoli, Brussels sprouts, cabbage, cauliflower, kale, kohlrabi, mustard, radishes, rutabagas, turnips), celery, corn, lettuce, nightshade family crops (eggplants, peppers, potatoes, tomatillos, tomatoes), onions, peas, sweet potatoes

Beet armyworm

Beet armyworm plant damage

SIGNS OF THEIR ACTIVITY

Skeletonized leaves of host plants; tunneling into the heads of lettuce, cabbage, or broccoli; small holes in tomato fruits; abnormal growth of leaves and flower stalks of artichokes; eggs on asparagus ferns; larvae feeding on asparagus ferns

NATURAL PREDATORS

Big-eyed bugs, damsel bugs, minute pirate bugs, tachinid flies

CONTROLS

> Practice crop rotation.

> Use row covers over small crops that don't need to be pollinated.

> Monitor plants frequently for damage.

> Hand-pick larvae, crush eggs.

> Apply *Bacillus thuringiensis* variety *kurstaki*, beneficial nematodes, Neem, plant extracts, pyrethrins, or spinosad at the first sign of damage.

> Cultivate the top 2 inches (5 cm) of the soil at the end of the season to unearth overwintering pupae.

BLISTER BEETLE

There are more than 2,500 species of blister beetles found worldwide. Two of the most common that cause damage to edible crops are margined blister beetles and striped blister beetles. Margined blister beetles have black wing covers with light gray edging while striped blister beetles are either orange or yellow with black stripes. All blister beetles are notable for having a rounded head that is wider than their thorax. Most have long, slender bodies that range from ⅜ to 1 inch (1 to 2.5 cm) in length and soft wing covers. Adults primarily eat flowers but many defoliate plants as well. When disturbed, blister beetles release a toxin called cantharidin, which is toxic to livestock if present in hay or other feed crops, and will cause blisters on human skin. The larvae are considered beneficials because they consume grasshopper eggs, but they also feed on the eggs of solitary bees.

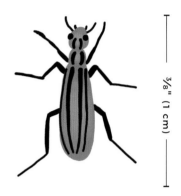

⅜" (1 cm)

Family name: Meloidae
Latin name: *Epicauta funebris* (margined), *Epicauta vittata* (striped)

Margined blister beetle

LIFE CYCLE

Blister beetles go through a complete metamorphosis from egg to larva to pupa to adult. The mature larvae overwinter in the soil and pupate in the spring, emerging as adults two weeks later. After mating, female beetles lay clusters of eggs in the soil, under rocks, or on host plants. The tiny first stage of a larva, referred to as a triungulin, feeds on grasshopper eggs and will even hitch a ride on a bee so it can eat the eggs in its nest. The larvae go through three instars.

TYPICALLY SEEN ON

Legume family crops (beans, peas, soybeans), nightshade family crops (eggplants, peppers, potatoes, tomatillos, tomatoes)

Striped blister beetle

Blister beetle damage on vegetable plant

SIGNS OF THEIR ACTIVITY

Chewed flowers and/or leaves

NATURAL PREDATORS

Birds

CONTROLS

> Monitor susceptible crops frequently.

> Hand-pick beetles (wear gloves).

> Sprinkle diatomaceous earth on and around host crops.

> Place floating row covers over any host crops that don't require pollination, such as potatoes.

> Keep up with your weeding since blister beetles are attracted to them.

> Attract birds to your garden since they can safely eat blister beetles.

> Apply pyrethrins or spinosad to plants but avoid spraying on or near flowers, as both products will kill bees.

CABBAGE LOOPER

Cabbage looper moth larvae primarily chew on the leaves of cabbage family crops but can be found on a few other vegetables, as well. While they are widespread, loopers only overwinter in regions where the temperatures don't drop below freezing. These caterpillars are pale green with white, lengthwise stripes and three pairs of prolegs that cause them to move in a looping motion typical of inchworms. They can grow up to 1½ inches (3.8 cm) long. Unlike cabbage worms (*Pieris rapae*, page 70), which are the larvae of butterflies, loopers have fairly smooth skin. The adult moths have speckled gray forewings with an unusual white pattern on them and pale brown hindwings; they are active at night.

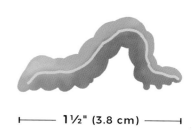

├─── 1½" (3.8 cm) ───┤

Family name: Noctuidae
Latin name: *Trichoplusia ni*

Cabbage looper moth

LIFE CYCLE

The moths lay round, cream-colored eggs that hatch in 2 to 5 days. The larvae go through several instars over the course of 3 weeks. After pupating in a thin, silky cocoon located on or near the same plant upon which they've fed, they emerge as moths in 1 to 2 weeks. Even though the adult stage of the moth only lasts about 12 days, they can lay up to 600 eggs in that time. Cabbage loopers produce multiple generations in a year.

TYPICALLY SEEN ON

Cabbage family crops (arugula, broccoli, Brussels sprouts, cabbage, cauliflower, kale, kohlrabi, mustard, radishes, rutabagas, turnips), lettuce, peas, potatoes, spinach

Cabbage looper caterpillar

Cabbage looper damage

SIGNS OF THEIR ACTIVITY

Feeding damage on the undersides of the leaves; jagged holes through the leaves; holes that bore into the heads of cabbage or broccoli; green frass around or below the plant damage

NATURAL PREDATORS

Big-eyed bugs, parasitic wasps, tachinid flies

CONTROLS

> Hand-pick the caterpillars, crush the eggs.

> Apply *Bacillus thuringiensis* variety *kurstaki* to the leaves when you detect damage.

> Apply insecticidal soap, Neem, plant extracts, pyrethrins, or spinosad.

> Clean up garden debris and eliminate nearby weedy areas during and at the end of the garden season to eliminate pupae.

CABBAGE WORM (IMPORTED)

Cabbage worms are the larvae of imported cabbage white butterflies. The adults have white wings with black tips and spots and a 2-inch (5 cm) wingspan. The females lay their ribbed eggs, which are creamy white to yellow, on the undersides of leaves. The caterpillars, which range from 1 to 1½ inches (2.5 to 3.8 cm) long, are pale green with a yellow line running down the center of their backs and are covered with fuzzy hairs. They also have 5 pairs of prolegs. The larvae chew on the leaves and can tunnel into the heads of cabbage and broccoli. The butterflies do not directly cause damage to the plants. While cabbage loopers (*Trichoplusia ni*) are a similar pest of the cabbage family, they are larvae of a moth; refer to their profile on page 68 for more information.

├─── 1½" (3.8 cm) ───┤

Family name: Pieridae
Latin name: *Pieris rapae*, also known as *Artogeia rapae*

Cabbage white butterfly

LIFE CYCLE

The butterfly emerges from its chrysalis in the spring, mates, and lays eggs on cabbage family crops. The larvae (caterpillars) hatch in 3 to 5 days. They go through five instars over the course of about 15 days, then pupate in a loose chrysalis that matches the color of the plant leaf or debris to which it is attached. In the summer, pupations take 11 days; otherwise, they pupate through the winter and emerge as butterflies in the spring to repeat the cycle. The butterflies live for 3 weeks and can lay up to 400 eggs.

TYPICALLY SEEN ON

Cabbage family crops (arugula, broccoli, Brussels sprouts, cabbage, cauliflower, kale, kohlrabi, mustard, radishes, rutabagas, turnips)

Cabbage worm

Large, irregular holes are common signs of cabbage worm activity.

SIGNS OF THEIR ACTIVITY

Irregular holes in the leaves; frass on and below the damaged plant; green caterpillars on the midrib of leaves

NATURAL PREDATORS

Assassin bugs, big-eyed bugs, damsel bugs, green lacewings, hover flies, parasitic wasps, spiders, syrphid flies, tachinid flies

CONTROLS

> Cover plants with floating row covers or tulle netting for the entire season since they don't require pollination.

> Hand-pick the larvae, crush the eggs.

> Spray *Bacillus thuringiensis* variety *kurstaki*, which is most effective on young caterpillars, on the leaves at the first sign of damage.

> Clean up plant debris to eliminate places for pupation to take place.

> Be sure to soak harvested crops in salty water, which will dislodge larvae and make them float to the surface.

> Other controls include Neem, plant extracts, pyrethrins, and spinosad.

CARROT RUST FLY

Carrot rust flies, which are also known as carrot root flies, are found in temperate climates. The adults are ⅕ inch (5 mm) long and have shiny black bodies, orange heads with red eyes, and orange legs. Their eggs are white. The larvae are cream-colored maggots with tapered bodies that are ¼ inch (6 mm) long. The scent of carrots and other carrot family crops, which are listed below, attract these flies. While they don't cause damage to plants, their larvae wreak havoc with the roots, causing plants to wilt or even die. Larval activity also results in the stunting and forking of the roots themselves, which can make crops such as carrots and parsnips unappealing.

¼" (6 mm)

Family name: Psilidae
Latin name: *Psila rosae*

Carrot rust fly

LIFE CYCLE

The fly overwinters, either as a larva in the soil or a pupa in the roots of the host plants, and pupates in early spring. They lay eggs on the soil at the base of the plants. After hatching, the maggots start by chewing on the carrot root hairs, then tunnel through the roots, leaving behind a rust-colored frass. The larvae go through three instars, then exit the carrot and pupate in the soil. There can be two to three generations in a year.

TYPICALLY SEEN ON

Carrot family crops (carrots, celeriac, celery, dill, fennel, parsnips)

Carrot rust fly larva and damage

SIGNS OF THEIR ACTIVITY

Tunnels and grooves in the host plants' roots, rust-colored frass; wilting or dead plants

NATURAL PREDATORS

Ground beetles, parasitic wasps, rove beetles

CONTROLS

> Practice crop rotation.

> Plant crops in early summer in cold winter areas or in late fall to late winter in regions with mild winters.

> Choose carrot varieties that are resistant to carrot rust flies, such as 'Flyaway'.

> To avoid attracting flies, disguise the scent of carrot family crops by intercropping with garlic or onions; space seeds to eliminate the need for thinning seedlings later as this also releases the carrot scent.

> Protect carrots and parsnips with row covers all season, securing all edges to prevent flies from gaining access.

> Since the adults don't fly higher than the plants' leaves, erect a 24-inch-high (61 cm) barrier around the planting (refer to the DIY project on page 152).

> Don't overwinter carrots in the ground.

> Clean up debris from carrot family crops at the end of the season to eliminate any overwintering larvae or pupae.

> Eliminate weeds that are members of the carrot family (Queen Anne's Lace, cow parsley).

COLORADO POTATO BEETLE

Colorado potato beetles are quite colorful, with 10 black-and-yellow stripes on their wing covers and red heads with black spots. They are about ½ inch (1.3 cm) long and ⅜ inch (1 cm) wide. The larvae have plump, rusty-red bodies with black heads, and two rows of black spots on each side of their body; they grow up to ½ inch (1.3 cm) long. The eggs are yellow to orange in color and laid on the undersides of the leaves. Both the adults and larvae chew on plant leaves but the latter are much more destructive. Once damage to plants exceeds 20 percent, it will affect overall yields and quality.

½" (1.3 cm)

Family name:
Chrysomelidae
Latin name: *Leptinotarsa decemlineata*

Colorado potato beetle

LIFE CYCLE

The adult beetles overwinter several inches (7.6 cm) down in the soil where a previous planting of a host crop grew. They emerge in the spring and, after mating, the females lay eggs on nightshade family plants. After 4 to 10 days, the larvae hatch and begin feeding on the plants' foliage. They go through four instars, then drop into the soil to pupate for up to 2 weeks before emerging as beetles. There can be as many as three generations per year.

TYPICALLY SEEN ON

Primarily potatoes; also other crops (eggplants, peppers, tomatillos, tomatoes) and weeds within the nightshade family

Colorado potato beetle larva

Colorado potato beetle eggs

SIGNS OF THEIR ACTIVITY

Chewed foliage, dark frass, yellow or orange eggs on the undersides of leaves

NATURAL PREDATORS

Assassin bugs, damsel bugs, ground beetles, lacewings, ladybugs, parasitic wasps, praying mantids, predatory stink bugs, robber flies, soldier beetles, spiders, tachinid flies

CONTROLS

> Colorado potato beetles have become resistant to many pesticides, which is a perfect example of why organic controls are so important to use.

> Practice a 3-year crop rotation.

> Consider "chitting" your seed potatoes (exposing them to light and allowing them to sprout) prior to planting; this gives them a jumpstart on growing so they can better withstand feeding damage.

> Monitor all nightshade crops regularly, including the undersides of leaves, for eggs.

> Hand-pick adult beetles and larvae; crush eggs.

> Cover potato plants with floating row covers for the entire season since they do not require pollination; for other nightshade crops, cover plants until they begin blooming.

> If Colorado potato beetles become a serious pest, consider growing nightshade family crops every other year to break the cycle.

> Apply beneficial nematodes, kaolin clay, Neem, plant extracts, pyrethrins, or spinosad.

MEET THE BUGS

CORN EARWORM

Corn earworms are also known as tomato fruitworms and cotton bollworm. But no matter what you call them, they are one of the most destructive agricultural pests. The adult moth lays eggs on host plants, such as corn when the silks begin developing (although they will also lay on dry brown silks), on the leaves near the developing fruit of tomatoes and peppers, and on lettuce leaves. The moth, which is nocturnal and feeds on nectar, is buff-colored, ¾ inch (2 cm) long and has a wingspan of 1 to 1½ inches (2.5 to 3.8 cm). They lay small white eggs with ridges. The 1- to 1½-inch (2.5 to 3.8 cm) larvae are pale but change to green or black as they mature. Corn earworms have a wide distribution. Even though they don't overwinter in colder climates, the moths will migrate north. Earworms can be confused with tobacco budworms but the latter aren't vegetable pests.

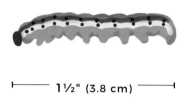

├─────── 1½" (3.8 cm) ───────┤

Family name: Noctuidae
Latin name: *Helicoverpa zea*

Corn earworm moth

LIFE CYCLE

While the female lays eggs singly on corn stalks, she also lay eggs on other areas of the plant and can lay up to 3,000 eggs in her short lifetime. The larvae hatch in 2 to 5 days and begin tunneling into the fruit. Initially, they feed together but as they mature, they become cannibalistic and eat one another. It is typical to find a single larva in an ear of corn. After feeding for about 4 weeks, during which each larva goes through several instars, it drops from the plant, creates a capsule, and pupates at a depth of 2 to 4 inches (5 to 10 cm) in the soil. It emerges about 10 days later to begin the cycle all over again.

TYPICALLY SEEN ON

Beans, corn, lettuce, peppers, tomatoes

Corn earworm caterpillar

Corn earworm damage to an ear of corn

SIGNS OF THEIR ACTIVITY

Frass at the tip of an ear of corn, silks chewed off, a worm in the tip of an ear, holes bored into bean pods, lettuce heads, peppers, tomatoes

NATURAL PREDATORS

Big-eyed bugs, damsel bugs, ground beetles, lacewings, ladybugs, minute pirate bug, parasitic wasps, spiders, tachinid flies

CONTROLS

> Plant corn earlier in the season if that is an option in your region.

> Choose early ripening varieties of corn.

> Monitor the plants regularly for damage.

> Apply a few drops of vegetable oil to young corn silks to smother eggs and young larvae.

> Apply *Bacillus thuringiensis* variety *kurstaki* (although it can be difficult to reach the larvae), beneficial nematodes, pyrethrins, or spinosad.

MEET THE BUGS

CUCUMBER BEETLE, STRIPED AND SPOTTED

It's bad enough when pests chew our plants, but when they carry deadly plant diseases as well, gardeners need to be extra vigilant. Because both of these cucumber beetles transmit bacterial wilt and cucumber mosaic virus, it is important to control them in your garden. The striped cucumber beetle is ⅕ inch (5 mm) long, yellow with three black stripes, and has a black head. The spotted cucumber beetle is ¼ inch (6 mm) long, yellow-green, has 12 black spots on its wings, and also has a black head. Both beetles cause more damage to plants than do their larvae. The beetles feed on young plants as well as the flowers, which prevents pollination and fruit-set, and on the fruits. The larvae are pale with black or brown heads and the eggs are orange to orange-yellow.

⅕" (5 mm)

Family name: Chrysomelidae
Latin name: *Acalymma vittatum* (striped); *Diabrotica undecimpunctata howardi Barber* (spotted)

Striped cucumber beetle

LIFE CYCLE

Both beetles overwinter in debris in the garden or close to it. They emerge in the spring and lay eggs on the ground near host plants. The larvae eat plant roots as well as the rind of fruits lying on the soil surface for about a month, then they pupate in the soil. There is an average of two generations per year.

TYPICALLY SEEN ON

Primarily cucurbit family crops (cucumbers, melons, pumpkins, squash); also attracted to asparagus, beans, beets, corn, potatoes, tomatoes

SIGNS OF THEIR ACTIVITY

Irregular holes in the leaves and flowers of host plants; chewed leaves of seedlings, wilting plants

NATURAL PREDATORS

Assassin bugs, parasitic wasps, tachinid flies

Spotted cucumber beetle

Cucumber beetle damage

CONTROLS

> Clean up garden debris at the end of the season.

> Practice crop rotation.

> Select cucurbit varieties that are resistant to bacterial wilt and cucumber mosaic virus.

> Start cucurbits indoors to let the seedlings become more vigorous before transplanting them outdoors so they can better withstand beetle damage.

> Try planting host plants later, if possible, to miss the peak feeding period of beetles.

> Add aromatic plants, such as marigolds or catnip, to the garden to repel the beetles

> Cover seedlings with floating row covers until plants start to bloom.

> Mulch heavily around the host plants to prevent beetles from laying their eggs in the soil.

> Place developing cucurbit fruits on boards or other supports so they're not in contact with the soil.

> Monitor daily for chewed leaves, flowers, or wilting; hand-pick beetles.

> Note that goldenrod and asters are alternate host plants so remove them if they are near your garden.

> Since allspice, clove, and bay oil are pheromones that attract the female beetles, make yellow sticky traps (refer to the DIY project on page 188) that contain cotton balls soaked in one of these botanical oils.

> Apply beneficial nematodes, kaolin clay, Neem, pyrethrins, or spinosad.

CUTWORM

There are many different species of cutworms. The adult moths are gray, brown, or black, with mottled or banded wings; they do not damage plants. The larvae, which are nocturnal, have smooth brown, tan, green, gray, or black bodies, with some having spots or stripes. They range from 1 to 2 inches (2.5 to 5 cm) in length and curl into a C-shape when disturbed. Cutworms favor seedlings with tender stems so are most active in the spring. While most gardeners are familiar with seeing the larvae in or on the soil, variegated cutworms (*Peridroma saucia*) will climb into plants, trees, and shrubs to feed on leaves, stems, and fruits.

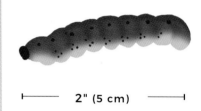

|← 2" (5 cm) →|

Family name: Noctuidae
Latin name: Many different species

Dingy cutworm moth (*Feltia jaculifera*)

LIFE CYCLE

During late summer and fall, the moths lay eggs on weeds, grasses, and plant debris or in the soil. After the larvae hatch, they feed on leaves or roots. They will remain active in areas with mild winters; in colder regions, they overwinter in the soil as larvae or pupae. In the spring, larvae curl around plant stems and chew through them. There are up to three generations in a year.

TYPICALLY SEEN ON

Artichokes, asparagus, beans, cabbage family crops (arugula, broccoli, Brussels sprouts, cabbage, cauliflower, kale, kohlrabi, mustard, radishes, rutabagas, turnips), carrots, celery, corn, cucurbit family crops (cucumbers, melons, pumpkins, squash), lettuce, peas, peppers, potatoes, tomatoes

Cutworms

Cutworm damage to an asparagus spear

SIGNS OF THEIR ACTIVITY

Seedlings or plant stems cut off at, just above, or just below the soil surface; cavities chewed into potato tubers; tips of asparagus spears deformed from chewing damage

NATURAL PREDATORS

Assassin bugs, damsel bugs, ground beetles, parasitic wasps, spiders, tachinid flies

SUSAN'S ANECDOTE

Several years ago, I expanded my garden into what had been a grassy area. The morning after transplanting melon seedlings into a new raised bed, I found a few of them lying on the ground, minus their roots. Cutworms were the guilty parties.

CONTROLS

> Monitor regularly for cutworms by looking for plant damage in the morning and by going out at night with a flashlight to spot them while they're active.

> Hand-pick the cutworms.

> Keep the garden tidy by removing plant debris and weeds to eliminate places for moths to lay eggs.

> Make barriers to protect seedlings (see the DIY project on page 160).

> Sprinkle diatomaceous earth around the base of the seedlings.

> Apply *Bacillus thuringiensis* variety *kurstaki*, beneficial nematodes, Neem, or plant extracts.

> If cutworms are an annual problem, disturb the soil in early spring and again in fall to expose any overwintering larvae or pupae.

MEET THE BUGS

DIAMONDBACK CATERPILLAR

Just like cabbage loopers and cabbage worms, diamondback moth larvae are damaging pests of cabbage family crops. The adult has a slender, ½-inch-long (1.3 cm) body, gray wings with golden diamond patterns, and long antennae. Larvae are small green caterpillars that grow up to ½ inch (1.3 cm) long by the time they reach their fourth and final instar. The pair of prolegs at the end of the body creates a distinctive V-pattern. When disturbed, the caterpillars will wiggle agitatedly and either drop off the plant or suspend themselves from a leaf by a silken strand. Only the caterpillars damage plants. The tiny eggs are yellow to pale green and difficult to see without a hand lens. Diamondback moths are found worldwide and produce up to a dozen generations each summer in warmer regions. They have a high mortality rate in areas with cold winters. Their population increases significantly when temperatures rise above 80°F (27°C).

½" (1.3 cm)

Family name: Plutellidae
Latin name: *Plutella xylostella*

Diamondback moth

LIFE CYCLE

The moths overwinter as adults in plant residue. In the spring, they mate and lay eggs, either singly or in small groups, on leaves. After hatching, the tiny larvae first engage in leaf-mining activities. While going through their four instars, they transition to feeding on the undersides of leaves. Once mature, they pupate in a loose silk cocoon on a leaf or in soil debris.

TYPICALLY SEEN ON

Cabbage family crops (arugula, broccoli, Brussels sprouts, cabbage, cauliflower, kale, kohlrabi, mustard, radishes, rutabagas, turnips)

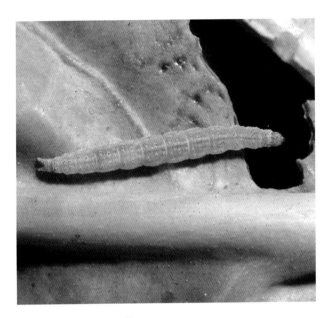
Diamondback caterpillar

Diamondback caterpillar damage to a plant

SIGNS OF THEIR ACTIVITY

Irregular holes in the leaves and stems; chewing damage to flower stalks and the heads of broccoli, Brussels sprouts, and cauliflower; stunted young seedlings with chewing damage to their growing points

NATURAL PREDATORS

Ground beetles, parasitic wasps, spiders, syrphid flies

CONTROLS

> Because diamondback moths and caterpillars are very resistant to pesticides, it's important to use organic control methods.

> Closely inspect cabbage family seedlings before purchasing them.

> Practice crop rotation.

> Use floating row covers as barriers to keep moths away from the plants.

> Monitor plants regularly for damage.

> Since larvae have a high mortality rate during rainstorms, consider using overhead watering for your cabbage family crops or occasionally hand-spraying them with water.

> Interplant cabbage family crops with other families of crops.

> Clean up crop residues at the end of the season.

> Apply *Bacillus thuringiensis* variety *kurstaki*, Neem, pyrethrins, or spinosad.

EARWIG

Their brown bodies and the pincers on their hind end make earwigs instantly recognizable. While not aggressive toward humans, they can pinch with their pincers. Even though they can cause trouble in the garden, they are considered beneficial due to their predation on aphids, mites, and other insects' eggs. Earwigs are nocturnal and hide in moist, dark areas, such as under garden debris or boards, during the daytime. Although there are approximately 1,800 species worldwide, only about 25 inhabit North America. Introduced to the continent in the early 20th century, European and ringlegged earwigs are common in US gardens. European earwigs (*Forficula auricularia*) are ½ to ¾ inch (1.3 to 2 cm) long and have wings underneath short wing covers, although they rarely fly. The roughly ½-inch-long (1.3 cm) ringlegged earwigs (*Euborellia annulipes*) prefer warmer regions, do not have wings, and are primarily predators.

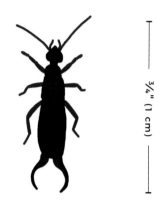

¾" (1 cm)

Family name: Forficulidae
Latin name: Many different species

Adult earwig

LIFE CYCLE

During the fall and winter months, female earwigs dig a chamber in the soil and lay an average of 30 eggs. The mother cares for the nymphs until they have gone through their first molt. She then creates an opening to the soil surface to allow the nymphs to leave the nest in search of food. They become independent after going through their third molt. Earwigs usually have one generation per year.

TYPICALLY SEEN ON

Artichokes, young bean seedlings, celery, corn ears, lettuce, peppers, potatoes, soft fruits

SIGNS OF THEIR ACTIVITY

Chewed corn silks, which impacts pollination and leads to poor ear development; frass; irregular holes in leaves; holes in peppers

Earwigs chew on the tips of ears of corn.

Earwig damage to celery stalks

NATURAL PREDATORS

Assassin bugs, birds, ducks, poultry, praying mantids, tachinid flies, toads

CONTROLS

> Keep your garden tidy by removing the hiding places that attract earwigs.

> Set earwig traps (refer to the DIY on page 162).

> Apply beneficial nematodes, diatomaceous earth, insecticidal soap, plant extracts, pyrethrins, or spinosad.

SUSAN'S ANECDOTE

The year 2019 was "the year of the earwig" in my garden and in many of my friends' gardens as well. The earwigs skeletonized my yardlong bean seedlings and chewed mercilessly on the celery and broccoli plants. I learned a valuable lesson that year: Earwigs will brutalize some plants if these pests can stay well-hidden from the gardener. I had covered my broccoli bed with a floating row cover to protect them from cabbage butterflies and aphids. At one point when I lifted off the cover to check on the plants, I was horrified to see scores of aphids falling off of them! I switched to a cover of bridal veil netting (tulle, page 145), which is much easier to see through. This made the earwigs feel more exposed and made them go elsewhere.

MEET THE BUGS

FLEA BEETLE

For being such small insects, flea beetles are one of the most challenging to control because they're so mobile. The adults are equipped with strong hind legs for jumping whenever they are disturbed, and they are able to fly long distances. Beetles range from ¹⁄₁₆ to ¼ inch (2 to 6 mm) long, are shiny black to brown in color, and some species have stripes. Both the eggs and larvae are creamy white. The adult stage is the most damaging, especially for young seedlings. Once the plants mature, they are less likely to be bothered. Even though there are many different species of leaf beetles, the damage caused by most looks the same: small, irregular "shot holes" throughout the leaves.

¼" (6 mm)

Family name:
Chrysomelidae
Latin name:
Many different species

Flea beetle

LIFE CYCLE

Adults lay eggs in the soil near the target plants. Once the larvae hatch, they eat plant roots, although their activity doesn't adversely affect plant health. The larvae pupate in the soil. After overwintering in plant debris, adult beetles move to their plant hosts in the spring. The number of generations per year varies with the species but the typical range is from one to three generations.

TYPICALLY SEEN ON

Artichoke, beet family crops (beets, spinach, Swiss chard), cabbage family crops (arugula, broccoli, Brussels sprouts, cabbage, cauliflower, kale, kohlrabi, mustard, radishes, rutabagas, turnips), corn, lettuce, nightshade family crops (eggplants, peppers, potatoes, tomatillos, tomatoes)

Flea beetle damage on eggplant leaf

Flea beetles on kale leaf

SIGNS OF THEIR ACTIVITY

Numerous small holes that are less than ⅛ inch (3 mm) in diameter in the leaves and stems of young plants; wilted, stunted seedlings; scars on potato tubers caused by the tuber flea beetle

NATURAL PREDATORS

Big-eyed bugs and parasitic wasps

CONTROLS

> Plant as late as you can to avoid peak flea beetle activity.

> Prior to planting, consider covering the soil with a reflective silver plastic mulch, making it more difficult for flea beetles to find host plants (refer to the DIY project on page 178).

> Monitor plants frequently for damage.

> Keep plants healthy and unstressed so they will be better able to survive beetle damage.

> Cover target crops with floating row covers (for plants that require pollination, remove the cover when they begin to bloom).

> Plant "trap crops," such as radishes, to divert flea beetles from more desirable crops.

> Place yellow sticky traps near target crops to determine if beetles are present (refer to the DIY project on page 188).

> Apply beneficial nematodes, diatomaceous earth, kaolin clay, Neem, plant extracts, pyrethrins, or spinosad.

> Control weeds in the garden and clean up garden debris at the end of the season.

MEET THE BUGS

GRASSHOPPER

Grasshoppers are easily recognizable insects that use their chewing mouthparts to feed on tender plant leaves. The adults grow up to 3 inches (7.6 cm) long and have strong hind legs and short antennae. They also have fully developed wings that allow them to fly great distances, making them difficult to control. The adults and nymphs have green, brown, or yellow coloration, often in combination. Grasshoppers are sometimes confused with katydids and crickets, which both have longer antennae.

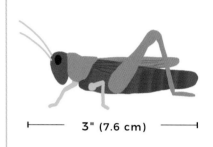

├─── **3" (7.6 cm)** ───┤

Family name: Acrididae
Latin name:
Many different species

Adult grasshopper

LIFE CYCLE

Grasshoppers breed in grassy areas, ditches, and fields in late summer through fall. The females lay up to 100 eggs, in clusters known as egg pods, in dry soil. The nymphs hatch in the spring, move to the surface, and begin feeding on plants. They go through several instars. Aside from being smaller and not yet having functioning wings, nymphs look very similar to the adults. Most species produce one generation per year. Swarms migrate to other areas when their food sources are scarce. Wet weather can significantly impact their population if it occurs while the eggs are hatching.

TYPICALLY SEEN ON

Beans, carrots, corn, leafy vegetables (such as lettuce, spinach), onions

Grasshopper nymph chewing holes in a leaf

Grasshoppers don't ordinarily bother squash plants but they made an exception in this case!

SIGNS OF THEIR ACTIVITY

While grasshoppers prefer to eat grasses, they can chew large, irregular holes into the leaves of vegetable and ornamental plants.

NATURAL PREDATORS

Birds, blister beetle larvae, coyotes, poultry, praying mantids, robber flies, soldier beetles

CONTROLS

> Add grasshopper-repelling plants, such as cilantro, calendula, and horehound, to the targeted areas of your garden.

> Protect susceptible crops with floating row covers; since grasshoppers can chew through lightweight row covers, you might need to switch to window screening.

> Attract birds to your garden to dine on grasshoppers.

> Apply insecticidal soap, kaolin clay, Neem, plant extracts, or pyrethrins while grasshoppers are young and more easily affected by them.

> If grasshoppers are particularly problematic, introduce the protozoan *Nosema locustae* while the larvae are in their first and second instars; it is available under the brand names of Nolo Bait and Semaspore and will not negatively impact other types of insects. Disturb the top 2 inches (5 cm) of soil at the end of the growing season to excavate the egg pods, making them vulnerable to predators and the elements.

HARLEQUIN BUG

It's a shame harlequin bugs are so destructive, because they sure are pretty. As a type of stink bug, the adults have ⅜ inch (1 cm) long, shield-shaped bodies and are very colorful, with yellow or orange-and-black patterns. They emit an unpleasant odor if disturbed. The nymphs have more rounded bodies with similar coloration and squiggly white stripes across their backs. Even the eggs are striking, barrel-shaped with black and white stripes. The adults and nymphs use their piercing/sucking mouthparts to extract sap from the host plants, which causes a cloudy discoloration of the foliage. Harlequin bugs are more prevalent in regions with milder winters although they have slowly been expanding their range into colder climates.

⅜" (1 cm)

Family name: Pentatomidae
Latin name: *Murgantia histrionica*

Harlequin bug adult

LIFE CYCLE

The adults overwinter in garden debris, emerge in the spring, and begin feeding on weeds within the cabbage family. In early summer, they feed on cabbage family crops. The females lay clusters of eggs, usually in rows of six. The nymphs hatch 1 to 2 weeks later and feed on the same host plant for 2 months. During this time, they go through five to six instars, getting larger each time they molt. Depending on the climate, there can be up to four generations per year.

TYPICALLY SEEN ON

Primarily cabbage family crops (arugula, broccoli, Brussels sprouts, cabbage, cauliflower, collards, kale, kohlrabi, mustard, radishes, turnips); also asparagus, cucurbit family crops (cucumbers, melons, pumpkins, squash), legumes (beans, peas), onions, nightshade family crops (eggplants, peppers, potatoes, tomatillos, tomatoes)

Harlequin bug nymph

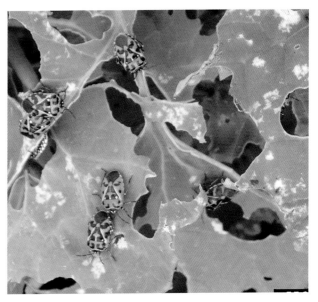
Harlequin bug damage on collard greens

SIGNS OF THEIR ACTIVITY

White blotches on leaves; discolored leaves; wilting leaves or plants; brown foliage

NATURAL PREDATORS

Big-eyed bugs, lacewings, ladybugs, minute pirate bugs, parasitic wasps, spiders (all prey on the eggs)

CONTROLS

> Look for vegetable cultivars (varieties) that are resistant to harlequin bugs.

> Monitor plants regularly, including leaf undersides, for all stages.

> Hand-pick adults and nymphs; crush eggs.

> Use floating row covers to protect crops that don't require pollination.

> Apply insecticidal soap, Neem, pyrethrins, or spinosad.

> If harlequin bugs have been a problem in your garden, dispose of host crop residues at the end of the season rather than composting them.

> Keep up with weeds both in and near your vegetable garden.

HORNWORM

Tomato hornworms are the caterpillar stage of the five-spotted hawk moth. These plump worms, which grow up to 4 inches (10 cm) long, are generally green but also can range from dark green to black. They have eight white, V-shaped lines on their backs and a black "horn" on their hind end. It is easy to confuse them with tobacco hornworms (*Manduca sexta*), which are the caterpillars of the Carolina sphinx moth; you can identify these worms by their seven diagonal white lines and red horn. Both species of hornworms feed on nightshade family crops. The five-spotted hawk moth has mottled, grayish-brown wings and five pairs of pale orange spots on the sides of their abdomens (the Carolina sphinx moth has six pairs). Hawk moths are nocturnal and usually observed fluttering like a hummingbird while feeding on the nectar of flowers. Their eggs are pale green. Since hornworms are well-camouflaged to blend in with the foliage, it can be difficult to spot them until they've caused a lot of damage.

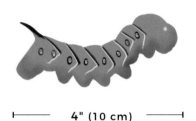

⊢———— **4" (10 cm)** ————⊣

Family name: Sphingidae
Latin name:
Manduca quinquemaculata
(tomato hornworm),
Manduca sexta
(tobacco hornworm)

Five-spotted hawk moth

LIFE CYCLE

Caterpillars overwinter as pupae in dark brown capsules several inches (7.6 cm) down in the soil. The adult moths emerge in spring and lay eggs singly on the foliage of host plants. The larvae hatch and go through several instars, during which they always have a horn. They mature within 3 to 4 weeks, then pupate for 2 weeks in the soil. They emerge in midsummer and begin the second generation.

TYPICALLY SEEN ON

Primarily tomatoes; other nightshade family crops (eggplants, peppers, potatoes, tomatillos)

Tobacco hornworms, shown here, are similar in appearance to tomato hornworms.

Hornworm damage on a tomato plant

SIGNS OF THEIR ACTIVITY

Chewed foliage and fruits; completely defoliated plants; dark frass on leaves and the ground below the feeding damage

NATURAL PREDATORS

Assassin bugs, lacewings, ladybugs, paper wasps, parasitic wasps

CONTROLS

> Remove weeds from the garden area as they can also host hornworms.

> Practice crop rotation.

> Monitor plants regularly for signs of damage or caterpillars.

> Hand-pick caterpillars.

> Apply *Bacillus thuringiensis* variety *kurstaki*, insecticidal soap, Neem, pyrethrins, or spinosad.

Note

If you spot a hornworm with small white cocoons hanging from its body, leave it alone. Those cocoons belong to parasitic wasps. The host will die and the emerging wasps will soon target more hornworms.

93

JAPANESE BEETLE

Japanese beetles have ½-inch-long (1.3 cm) metallic green bodies with bronze-colored wing covers. They have short, clubbed antennae as well as five tufts of white hairs along each side and two tufts at the tip of the abdomen. When disturbed, they quickly drop off plants. The adults feed on ornamental plants and vegetable crops. Beetles feed in groups, especially in hot weather, eating leaf tissue between the veins. This causes plants to release pheromones that attract even more beetles. The larvae are translucent, creamy-white grubs with dark heads and a V-shaped pattern of hairs on their hind end. They feed on the roots of turfgrass and are 1 inch (2.5 cm) in length when mature.

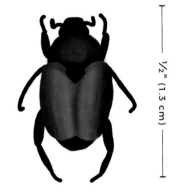

½" (1.3 cm)

Family name: Scarabaeidae
Latin name: *Popillia japonica*

Japanese beetle adults

LIFE CYCLE

Adult Japanese beetles emerge from the soil in June and move to their host plants where they feed and mate. During the summer, females periodically lay small clusters of eggs in the soil beneath turfgrass at a depth of 2 to 4 inches (5 to 10 cm). After hatching, the larvae feed on grass roots. They mature in a few weeks and move deeper into the soil to overwinter as pupae in cells. Beetles live 4 to 8 weeks and can lay up to 60 eggs during that time. The entire life cycle takes 1 year.

TYPICALLY SEEN ON

Basil, beans, collards, corn, eggplant, leafy greens (such as lettuce, spinach), okra, peppers, soybeans, tomatoes

Skeletonized leaves are a common sign of Japanese beetle activity.

Japanese beetle grubs

SIGNS OF THEIR ACTIVITY

Skeletonized leaves; chewed corn silks; dead sections of lawn

NATURAL PREDATORS

Assassin bugs, birds, moles, parasitic wasps, robber flies, skunks, tachinid flies

CONTROLS

> Be diligent about hand-picking the beetles early on to curb population levels. This will cut down on plant damage, which reduces the chemical signals that attract beetles.

> Apply kaolin clay, Neem, or pyrethrins for the beetles; apply beneficial nematodes to kill the grubs; apply milky spore (*Paenibacillus popilliae*, formerly known as *Bacillus popilliae*) right after egg-laying to infect and kill the grubs; or apply *Bacillus thuringiensis* variety *galleriae* for control of adults and grubs.

> Mow your lawn less frequently to keep it at a higher length; this increases lawn vigor, making it more tolerant of grub damage and less appealing for the adults to lay eggs.

> Because young grubs are very susceptible to dry conditions, cut back on the frequency of watering your lawn—or stop altogether—in summer.

> Do not use Japanese beetle traps as a means of control; research has found the floral lures contained in them attract more beetles to the area.

LEAFHOPPER, BEET AND POTATO

Leafhoppers are tiny, wedge-shaped insects that range between ⅛ to ³⁄₁₆ inch (3 to 5 mm) long. They use their piercing-sucking mouthparts to extract sap from plant leaves and stems. Beet leafhopper adults are light brown and potato leafhopper adults are pale green; both have wings. Their large hind legs enable them to hop from one plant to another. They also excrete sticky honeydew. The nymphs are paler than their adult counterparts and walk sideways like crabs. Both the adults and nymphs shed their skins on the leaf undersides, which is a good way to determine if you're dealing with leafhoppers. Beet leafhoppers are vectors for beet curly-top virus: They inject a pathogen into the plants that causes leaves to turn yellow and leaf veins to become purple. The virus also causes deformed plants and fruits.

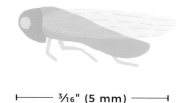

├────── ³⁄₁₆" (5 mm) ──────┤

Family name: Cicadellidae
Latin name: *Empoasca fabae* (potato), *Eutettix tenellus* (beet)

Beet leafhopper

LIFE CYCLE

Both leafhopper adults and eggs overwinter in sheltered areas. After the eggs hatch in the spring, the nymphs go through four to five instars over the course of 2 to 3 months. In the later stages, they develop wings; there is no pupal stage. Leafhoppers can produce two or more generations in a year. Some will die over the winter if it is particularly harsh. Adults are known to migrate long distances.

TYPICALLY SEEN ON

Beans, eggplants, potatoes (potato leafhoppers); beans, beets, peppers, tomatoes (beet leafhoppers)

Potato leafhopper

Leafhopper damage to fava bean leaves (note mottled appearance)

SIGNS OF THEIR ACTIVITY

White spots on the leaves; yellow leaves (this is known as "hopperburn"); curling leaves; distorted leaves; purple veins; stunted, deformed plants and fruits

NATURAL PREDATORS

Assassin bugs, big-eyed bugs, damsel bugs, lacewings, ladybugs, minute pirate bugs, praying mantids, predatory mites, robber flies, spiders, syrphid flies

CONTROLS

> Because they are so mobile, leafhoppers are difficult to control.

> Place row covers over host crops and remove them when the plants begin blooming to allow for pollination.

> Apply diatomaceous earth, insecticidal soaps, kaolin clay, Neem, plant extracts, or pyrethrins, especially to the undersides of the leaves.

> Dispose of diseased plants from your garden.

LEAFMINER

Many species of leafminers are pests of ornamental plants, but we'll limit our focus to those that impact vegetable crops. The name of this insect stems from the way the ¼ inch long (6 mm) maggots tunnel between the top and bottom layers of plant leaves as they feed. Gardeners rarely see the maggots themselves but it's easy to spot squiggly lines in the leaves or what appears to be a clear "window" in a leaf. The adult flies range from ⅕ to ⅓ inch (5 to 8 mm) long and are either black and yellow or have grayish bodies, depending on the species. The insects' activity rarely impacts the host plant's ability to grow but they can certainly ruin leafy crops, such as spinach and Swiss chard. The activity of asparagus leafminers is slightly different in that they tunnel through the base of asparagus stems. The leafminers listed at right have a wide distribution in many different climates.

¼" (6 mm)

Family name: Agromyzidae

Latin name: *Ophiomyia simplex* (Asparagus leafminer), *Pegomya betae* (beet leafminer), *Liriomyza brassicae* (cabbage leafminer), *Liromyza huidobrensis* (pea leafminer), *Pegomya hyoscyami* (spinach leafminer), *Liriomyza sativae* (vegetable/serpentine leafminer)

Spinach leafminer fly

LIFE CYCLE

After overwintering in the soil, the pupae emerge as flies in the spring to mate and lay their white, cylindrical eggs on the undersides of the leaves, usually in neat rows. The light-colored maggots hatch in 3 to 6 days and tunnel into the leaves, where they feed for up to 3 weeks. When mature, they cut a hole through the leaf and drop to the ground to pupate in the soil for 2 to 4 weeks. There can be several generations in a growing season.

TYPICALLY SEEN ON

Artichokes, beans, beet family crops (beets, spinach, Swiss chard), melons, peas, peppers, potatoes, squash, tomatoes

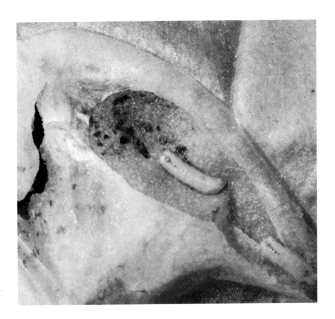

Leafminer larvae

Leafminer damage on spinach leaves

SIGNS OF THEIR ACTIVITY

Squiggly lines, blotchy areas, or clear patches in the leaves, as well as frass inside the leaves

NATURAL PREDATORS

Parasitic wasps, spiders

CONTROLS

> Practice crop rotation to make it difficult for overwintering pupae to find their host plants.

> Monitor plants regularly (especially on leaf undersides), crush eggs, and remove damaged leaves immediately.

> Avoid overwintering beet family crops.

> Clean up crop residue and debris at the end of the season.

> Weed regularly.

> Apply beneficial nematodes, Neem, plant extracts, pyrethrins, or spinosad to all leaf surfaces.

LYGUS BUG

Lygus bugs are particularly worrisome for vegetable gardeners due to the sheer number of crops they damage. The two most common species that target vegetable plants are the tarnished plant bug (*Lygus lineolaris*) and the Western tarnished plant bug (*Lygus hesperus*). The adults are ¼ inch (6 mm) long and ⅛ inch (3 mm) wide, green to brown in color, and have a distinctive V-shaped or triangular marking on their backs. Nymphs are light green and very similar in appearance to aphids but they move much more quickly and have red tips on their antennae. Eggs are cream-colored. Both the adults and nymphs have piercing-sucking mouthparts, which they use to extract plant sap. They can transmit plant diseases.

¼" (6 mm)

Family name: Miridae
Latin name: *Lygus* spp.
(Many species)

This is an adult Western tarnished plant bug, a species of lygus bug.

LIFE CYCLE

Adults overwinter in grassy areas or garden debris. In the spring, they emerge and usually feed on weeds. After mating, the female inserts eggs into the leaves and stems of broadleaf plants. The nymphs hatch 1 to 4 weeks later, depending on the weather conditions, and go through five instars. There can be up to four generations per year, depending on the climate.

TYPICALLY SEEN ON

Beet family crops (beets, spinach, Swiss chard), cabbage family crops (arugula, broccoli, Brussels sprouts, cabbage, cauliflower, kale, kohlrabi, mustard, radishes, rutabagas, turnips), carrot family crops (carrots, celery, parsnips), cucurbit family crops (cucumbers, melons, pumpkins, squash, watermelons), legume family crops (beans, peas), nightshade family crops (eggplants, peppers, potatoes, tomatillos, tomatoes)

Lygus bug damage to the seeds in a fava bean pod

SIGNS OF THEIR ACTIVITY

Discoloration of plant stems, leaves, and flowers; distorted plant growth; flowers and fruits of nightshade family crops drop prematurely; "cat-facing" damage (scarring and cavities) on plant fruits; discoloration and scarring of lettuce leaves; brown, wilting areas of celery; pitted bean pods

NATURAL PREDATORS

Assassin bugs, ground beetles, big-eyed bugs, damsel bugs, ladybugs, minute pirate bugs, predatory stink bugs, parasitic wasps, spiders

CONTROLS

> Monitor plants regularly for damage.

> Weed control in and around the vegetable garden is a must.

> Clean up garden debris at the end of the season.

> Apply insecticidal soap, kaolin clay, Neem, plant extracts, or pyrethrins.

MEXICAN BEAN BEETLE

Mexican bean beetles are a species of lady beetle that is not beneficial in the least. The ¼-inch-long (6 mm) adults look like large, coppery-orange ladybugs and have 16 black spots arranged in 3 rows on their wing covers. They are strong fliers that can cover long distances, making them difficult to control. Their yellow larvae have soft bodies covered with spines; the eggs are yellow as well.

¼" (6 mm)

Family name: Coccinellidae
Latin name: *Epilachna varivestis*

Mexican bean beetle adult

LIFE CYCLE

Adults overwinter in garden debris or in wooded areas, then emerge in the spring. Females lay clusters of yellow eggs on the undersides of the host plants' leaves. The larvae feed for about 3 weeks, remaining on the lower side of the leaves while going through four instars. They will also pupate in this location. Adults feed on the undersides of the leaves but will also chew on the bean flowers, pods, and stems. The peak of this damage occurs during the summer. There are typically two to three generations in a season.

TYPICALLY SEEN ON

Beans of all kinds but especially lima and snap beans, cowpeas and soybeans; occasionally on cabbage, kale, mustard

SIGNS OF THEIR ACTIVITY

Skeletonized leaves that can look transparent; leaves dropping off the plants; chewed bean pods

Mexican bean beetle larvae

Mexican bean beetle damage

NATURAL PREDATORS

Assassin bugs, ladybugs, minute pirate bugs, parasitic wasps, praying mantids, robber flies, spined soldier bugs, tachinid flies

CONTROLS

> If Mexican bean beetles are especially bad, consider skipping a year of growing their host plants.

> Look for resistant varieties.

> Choose short-season varieties to limit the damage.

> Prior to planting, consider covering the soil with a reflective silver plastic mulch, which will make it more difficult for beetles to find host plants (refer to the DIY on page 178).

> Use floating row covers as soon as you plant; remove them when the plants begin blooming to allow for pollination.

> Consider planting an early "trap crop" of beans to attract the beetles, then pull up and dispose of the plants in bags to remove the pests.

> Monitor plants frequently and crush any eggs, larvae, or adults you find.

> Clean up garden debris at the end of the season.

> It is possible to compost plant material if you first seal damaged plants in a plastic bag for a week to kill the beetles, larvae, and eggs.

> Apply kaolin clay, Neem, plant extracts, pyrethrins, or spinosad.

PILLBUG AND SOWBUG

Pillbugs and sowbugs are terrestrial crustaceans distantly related to shrimp and lobster. Pillbugs are also known as "roly-polies" for their ability to roll into a ball to protect themselves. Sowbugs have a more flattened appearance and two tail-like appendages on their hind ends that prevent them from rolling over. Both are also referred to as woodlice because they're usually found under logs. They are gray to brown with shell-like segmented bodies, have 7 pairs of legs, and breathe through gills on their abdomen. Adults are ¼ to ½ inch (0.6 to 1.3 cm) long. Primarily scavengers, they feed on decomposing plant material but can also damage young seedlings. Pillbugs and sowbugs are mostly nocturnal and hide in cool, moist areas under leaf litter, rocks, and boards. They don't sting or bite.

¼" (6 mm)

Family name: Armadillidiidae (pillbugs), Oniscidae (sowbugs)
Latin name: *Armadillidium vulgare* (pillbugs); *Porcellio scaber*, *P. laevis*, *Oniscus asellus* (sowbugs)

Pillbugs

LIFE CYCLE

The females deposit up to 200 eggs in a special pouch (aptly named the "marsupium"). Juveniles hatch in a month but remain in the pouch, where they feed on a special fluid. After a couple of weeks, they leave the pouch and go through several molts. It takes the young about a year to reach maturity. Because moisture is crucial to their existence, a large percentage of the young don't survive. Females can produce up to three broods per year. Pillbugs and sowbugs live from 2 to 5 years.

TYPICALLY SEEN ON

Seedlings of beans, cucumbers, lettuce, melons, mustards, onions, peas, peppers, squash, tomatoes

Sowbug

Pillbug damage to pumpkin seedling's stem

SIGNS OF THEIR ACTIVITY

Chewed stems and leaves of tender seedlings

NATURAL PREDATORS

Frogs, spiders, toads

SUSAN'S ANECDOTE

In 2017, after happily planting out bean and melon seedlings that I'd started indoors, I made an unhappy discovery the next morning: Some plants were lying on their sides, with their stems nearly chewed all the way through. I figured it was slugs, but after venturing out with a flashlight that night, I found pillbugs munching away on more stems. A sprinkling of diatomaceous earth around the remaining seedlings saved the day but I decided pillbugs aren't completely benign!

CONTROLS

> Keep garden free of debris under which they can hide.

> Apply diatomaceous earth around the base of susceptible seedlings to create a barrier.

> Hand-pick them in areas where they are a problem.

> Water in the morning so the garden can dry out before evening.

> Avoid excessive watering.

> Apply beneficial nematodes, diatomaceous earth, spinosad.

ROOT MAGGOT, CABBAGE AND ONION

There are several species of root maggots but the most damaging for vegetable crops are cabbage and onion maggots. The adult flies of both species are ¼ inch (6 mm) long and look like striped houseflies; the onion maggot fly has a humped back. Larvae are ¼ inch (6 mm) long, creamy white, and have pointed heads. They tunnel their way through the roots of cabbage and onion family crops and sometimes through stems. Their activity causes root rot and increases disease susceptibility. Root maggots are more prevalent during wet, cool springs. During periods of drought, they feed on roots above the soil surface. Some maggots damage germinating seeds.

¼" (6 mm)

Family name: Anthomyiidae
Latin name: *Delia radicum* (cabbage maggot), *D. antiqua* (onion maggot)

Onion maggot fly

LIFE CYCLE

The pupae overwinter in soil at a depth of 1 to 2 inches (2.5 to 5 cm). They emerge as adults in early to mid-spring, mate, and lay up to 200 eggs in cracks in the soil around the base of host plants. Maggots hatch in 10 days and feed on plant roots for 3 to 4 weeks. When mature, they move into the soil to pupate. The complete life cycle takes about 2 months and there can be up to three generations per year.

TYPICALLY SEEN ON

Beets, cabbage family crops (arugula, broccoli, Brussels sprouts, cabbage, cauliflower, kale, kohlrabi, mustard, radishes, rutabagas, turnips), carrots, onion family crops (garlic, leeks, onions, shallots)

SIGNS OF THEIR ACTIVITY

Cabbage family seedlings are wilted, stunted, have yellow leaves or die suddenly. The surfaces of root crops are scarred and there are tunnels through the roots.

Onion maggots and damage

Cabbage maggot damage to rutabaga

NATURAL PREDATORS

Ants, ground beetles, parasitic wasps, rove beetles

CONTROLS

> Prevention is vital, because once root maggots have burrowed into the host, it's too late.

> If root maggots were a problem in your garden last year, skip a year of growing the host crops.

> Practice crop rotation.

> Avoid using animal manure or decomposing plants in the soil in spring because these attract root maggots.

> Plant target crops in June to avoid peak egg-laying season.

> Place floating row covers over susceptible crops or make 4-inch-wide (10 cm) collars around the base of each seedling.

> Sprinkle wood ashes or diatomaceous earth around the base of target plants.

> Monitor target crops regularly for signs of damage or maggots.

> Apply pyrethrins.

> Clean up garden debris; destroy damaged crops, including the root systems.

> Add beneficial nematodes (*Heterorhabditis* or *Steinernemena*) annually.

SLUG AND SNAIL

With so many species of snails and slugs found worldwide, it's small wonder some of them make it into our gardens. Only snails have shells, but a strong, muscular foot is a characteristic shared by all species of gastropod—the name means "stomach foot." The slimy mucous they all secrete protects them and allows them to slide around their habitat. Both have two sets of tentacles: one pair for smell and the other, longer pair for vision. Garden slugs grow up to 2 inches (5 cm) long; snails have a similar length and shells that are 1 to 1¼ inches (2.5 to 3.2 cm) in diameter. They are active at night and hide in moist locations during the day. Slugs and snails use a rasping action to chew on plant leaves.

2" (5 cm)

Family name:
Agriolimacidae (gray garden slug); Helicidae (brown garden snail); all snails and slugs are in the Gastropoda class.
Latin name:
Deroceras reticulatum (gray garden slug), *Cornu aspersum* (brown garden snail)

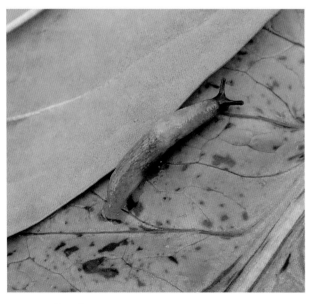

Gray garden slug

LIFE CYCLE

All snails and slugs are hermaphrodites, meaning they have both sexual reproductive organs. They lay 80 to 100 eggs in the spring and fall, in soil cracks or under leaves. Slugs reach full size in 3 to 6 months and snails mature in 2 years. Both hibernate in topsoil during the colder months of the year. During hot weather, snails estivate (hibernate) by surrounding themselves with a papery membrane. Unless they meet up with a duck or a cranky gardener, slugs live about 1 year and snails up to 6 years.

TYPICALLY SEEN ON

Artichokes, basil, beans, beet family crops (beets, spinach, Swiss chard), cabbage family crops (arugula, broccoli, Brussels sprouts, cabbage, cauliflower, kale, kohlrabi, mustard, radishes, rutabagas, turnips), lettuce, peppers, tomatoes

Brown garden snail

Classic slug damage on a lettuce leaf

SIGNS OF THEIR ACTIVITY

Irregularly-shaped holes in leaves; their slime trails are a dead giveaway, but do not confuse those with honeydew.

NATURAL PREDATORS

Ground beetles, rove beetles, ducks, geese, chickens, frogs, lizards

CONTROLS

> Eliminate moist hiding places; change from overhead watering to drip irrigation in problematic areas and water in early morning.

> Hand-pick them.

> Add aromatic plants, such as rosemary, lavender, sage, and nasturtiums, to repel them.

> Trap them under boards and upside-down flower pots.

> Use beer traps (refer to the DIY project on page 148) Create barriers with copper tape or flashing (refer to the DIY project on page 154).

> Apply diatomaceous earth or plant extracts.

> Use organic slug and snail baits that contain iron phosphate (avoid baits containing metaldehyde as it is toxic to dogs and cats).

SPIDER MITE

Just like spiders and ticks, spider mites are arachnids, not insects. They have tiny bodies (less than 1 mm long) with eight legs, and live in colonies. The two-spotted spider mite (*Tetranychus urticae*) is most commonly associated with damage to vegetable crops and is considered the most damaging of all spider mites around the world. It has a pale, yellow body with two dark spots and uses its piercing/sucking mouthparts to extract sap from plant leaves. They are most active during warm, dry spells.

Family name: Tetranychidae
Latin name:
Many different species

Spider mites

LIFE CYCLE

Both eggs and adult spider mites overwinter in the debris of host plants or other protected areas. The mites begin feeding on plants in the spring and soon lay eggs within fine webbing near the leaf veins. The eggs hatch into six-legged larvae that go through two immature stages before becoming eight-legged adults. This all takes place over 1 to 3 weeks, depending on the weather. Once she's an adult, a single female spider mite can lay hundreds of eggs. During high summer temperatures, colony populations increase dramatically as each new generation runs its course in one week. Some spider mites reproduce asexually. In mild climates, they are active year-round.

TYPICALLY SEEN ON

Beans, cucumbers, lettuce, melons, peas, tomatoes

Spider mite damage on a tomato leaf

Close up of two-spotted spider mites

SIGNS OF THEIR ACTIVITY

White or yellow stippling of the leaves; stunted growth; dying plants; fine webbing

NATURAL PREDATORS

Big-eyed bugs, damsel bugs, lacewings, ladybugs, minute pirate bugs, predatory mites, predatory thrips, spiders

CONTROLS

> Spider mites are more common in landscapes where pesticides are frequently used, which kills off their many predators. This underscores the importance of using organic control methods.

> Prior to planting, cover the soil with a reflective silver plastic mulch, which will make it more difficult for spider mites to find host plants (refer to the DIY on page 178).

> Monitor plants frequently, especially during hot, dry weather. The mites are so tiny they are hard to spot, so watch for their webbing.

> It is important to diagnose the problem correctly because the plant damage could also be as a result of drought stress. To determine if spider mites are the culprits, hold a sheet of paper under the leaves and shake them; some mites will drop down if they are present.

> Water plants regularly, and increase the amount they receive when summer temperatures heat up.

> Use a strong jet of water to dislodge mites from the affected plants.

> Apply horticultural oil, kaolin clay, plant extracts, pyrethrins, or spinosad.

MEET THE BUGS

SQUASH BUG

Adult squash bugs have dark brown or gray flattened bodies with orange-and-brown-striped abdomens; they are ⅝ inch (1.2 cm) long and ⅓ inch (8 mm) wide. Their eggs, which are reddish-brown and elliptical, hatch into nymphs that range from pale green to gray through subsequent molts. The adults and nymphs have piercing/sucking mouthparts that they use to remove sap from the plants. Squash bugs are carriers of a bacterium (*Serratia marcescens*) that causes cucurbit yellow vine disease; the leaves of infected plants all turn yellow over the course of a few days. Note that stink bugs have a similar appearance to the adults but do not target members of the cucurbit family.

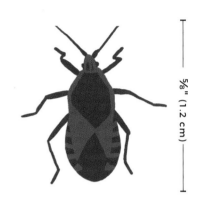

⅝" (1.2 cm)

Family name: Coreidae
Latin name: *Anasa tristis*

Adult squash bug

LIFE CYCLE

The adults live in garden debris through the winter months, then fly to young cucurbit plants in the spring to feed and mate. The female lays about 20 reddish-brown eggs on the undersides of the leaves in between the veins or on the plant stems. After roughly 10 days, the eggs hatch into small black, spider-like nymphs, which go through a total of five instars over the course of 4 to 6 weeks. In northern regions there is a single generation in a year, but in warmer areas there can be two to three generations.

TYPICALLY SEEN ON

Primarily pumpkins and winter squash; also other cucurbit family members (cucumbers, melons, summer squash)

SIGNS OF THEIR ACTIVITY

Tiny specks on leaves; leaves turn yellow, then brown, and start wilting; plant dies; reddish-brown eggs on stems or the undersides of leaves; nymphs and adults scurry away when you get close

Squash bugs typically lay their eggs in between the veins of leaves.

Squash bug nymphs go through five instars (developmental stages).

NATURAL PREDATORS

Praying mantis, spiders, tachinid flies

CONTROLS

> Use row covers as soon as you plant the seeds or seedlings and leave them on until plants begin blooming.

> Avoid using mulch under plants as it provides hiding places for adults and nymphs.

> Dispose of (don't compost) any plants exhibiting signs of cucurbit yellow vine disease.

> Practice good sanitation in the garden by cleaning up debris during and at the end of the season.

> Monitor the plants for damage or signs of the insects. Hand-pick nymphs and adults; crush eggs.

> Place boards under the plants, check under them in the morning, and destroy any nymphs or adults you find.

> Use crop rotation so cucurbits aren't planted in same place every year.

> Consider growing cucurbits vertically on trellises.

> Plant resistant cultivars, such as 'Butternut', 'Early Summer Crookneck', 'Improved Green Hubbard', 'Royal Acorn'.

> Apply kaolin clay, Neem, or pyrethrins.

SQUASH VINE BORER

The adult squash vine borer is a red-and-black moth with metallic green forewings and clear hindwings. They are 1½ inches (3.8 cm) long, look like a wasp, have a wingspan of 1½ inches (3.8 cm), and are active during the day. The larvae, which are fat, cream-colored caterpillars with brown heads, hatch from brown eggs and grow up to 1 inch (2.5 cm) long. Squash vine borers target cucurbit family crops but rarely bother cucumbers, melons, or watermelons. Larvae feed inside the stems, blocking the flow of water and nutrients throughout the vines, resulting in wilting and plant death.

1½" (3.8 cm)

Family name: Sesiidae
Latin name: *Melittia cucurbitae*

Squash vine borer moth

LIFE CYCLE

Pupae overwinter in brown cocoons within the soil where the previous year's host crops grew. The female moth emerges, mates, and lays eggs singly at the base of host plants or on the soil surface. After hatching a week later, larvae begin boring into the stems. They feed for a month, then exit the plant and pupate in the soil. There are one to two generations per year.

TYPICALLY SEEN ON

Pumpkins, summer and winter squash

SIGNS OF THEIR ACTIVITY

Yellowing leaves; wilting plants; puncture holes in stems near the base of the plant; golden frass; dead vines

NATURAL PREDATORS

Ground beetles, parasitic wasps

Squash vine borer larva

Squash vine borer damage

CONTROLS

> Practice a 3-year crop rotation of cucurbit crops.

> Consider planting a "trap crop" of squash and then disposing of it to eliminate eggs and larvae.

> Watch for moths flying near host plants. Since the color yellow attracts them, set out a yellow pail filled with water: They will fly into it and drown.

> Cover plants with floating row covers at planting time and anchor them well to prevent moths from getting under them; remove when plants begin flowering to permit pollination.

> To prevent egg-laying, wrap the lowest section of each plant's main stem with a collar of aluminum foil; periodically check and readjust or replace them with wider strips as each plant grows.

> Monitor plants for eggs; crush them.

> Apply diatomaceous earth or kaolin clay before the eggs hatch.

> Cut into the affected stem and remove or pierce through the larvae; immediately cover the area with soil or tape to help the plant heal.

> Inject beneficial nematodes (*Heterorhabditis bacteriophora, Steinernema carpocapsae*) into vines.

> Always destroy (don't compost) host plants with vine borer damage.

> Cultivate the top 2 inches (5 cm) of soil in fall to expose pupae to winter temperatures and predators.

STINK BUG

Stink bugs, sometimes called shield bugs for their shield-shaped bodies, are named for the unpleasant odor they emit as a defense mechanism when disturbed. Herbivorous stink bugs that bother vegetable crops include the Southern green stink bug (*Nezara viridula*) and the brown marmorated stink bug (*Halyomorpha halys* Stal). There are also predatory stink bugs (see page 129), which gardeners should welcome. Adult Southern green stink bugs have green ½-inch-long (1.3 cm) bodies. Brown marmorated adults are ⅝ inch (1.6 cm) long and have brown wing covers with an alternating brown-and-white pattern around the edge, plus two white bands on their antennae. Both species have scent glands on their thorax and abdomen. Each lay barrel-shaped eggs. The nymphs are often difficult to identify due to their wide color variations. Both adults and nymphs inject enzymes into vegetables and fruits that cause tissue damage. It is common to see the adults on the sides of your home or inside your house as they look for a place to hibernate over the winter. They are harmless to humans and do not sting or bite.

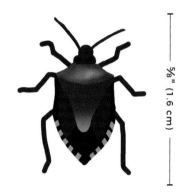

⅝" (1.6 cm)

Family name: Pentatomidae
Latin name:
Many different species, including *Nezara viridula* (Southern green) and *Halyomorpha halys* Stal (brown marmorated, shown here)

Brown marmorated stink bugs

LIFE CYCLE

Adult Southern green stink bugs and adult brown marmorated stink bugs emerge in the spring, mate, and the females lay masses of eggs on the undersides of leaves. After hatching, the nymphs go through five instars with associated color changes as they mature. They feed in groups initially but eventually disperse. In the fall, adults look for a place to overwinter. The entire life cycle takes from 40 to 70 days and there can be up to five generations in a year.

TYPICALLY SEEN ON

Primarily artichokes, beans, corn, soybeans, tomatoes, but also found on peas, peppers, squash

A Southern green stinkbug

Stink bug damage on artichokes

SIGNS OF THEIR ACTIVITY

Holes in artichoke heads and dark, damaged scales; frass; mottled skin of fruits; dimpled or otherwise distorted fruits or plants; fruits fall off plants; sunken skin; sunken bean seeds and pods

NATURAL PREDATORS

Assassin bugs, birds, frogs, parasitic wasps, praying mantids, predatory stink bugs, spiders

CONTROLS

› Place stink bug traps that employ pheromones to attract both male and female stink bugs in your garden.

› Hand-pick adults and nymphs; crush eggs.

› Use floating row covers over targeted crops but remove when plants begin blooming to allow for pollination.

› Apply insecticidal soap labeled for use on stink bugs, kaolin clay, or pyrethrins (use outdoors only).

› Seal up cracks and other areas where stink bugs can gain entry in the fall.

THRIPS

Thrips are tiny insects that use their piercing/sucking mouthparts to extract sap from the cells of plants. They can be a challenge to control since they are usually inside blossoms or emerging leaves. Bean thrips (*Caliothrips fasciatus*), onion thrips (*Thrips tabaci*), and western flower thrips (*Frankliniella occidentalis*) are common in vegetable gardens. Onion thrips can arrive at a garden on infested onion plant starts or onion sets (bulbs). Adult thrips are ¹⁄₁₆ inch (2 mm) long and have fringed wings. They range in color from yellow to brown or black. Thrips can transmit tomato spotted wilt virus, among other viral diseases. The adults are able to fly long distances, especially on windy days. In case you're wondering, the word "thrips" is both singular and plural.

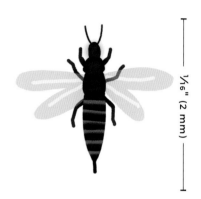

1⁄16" (2 mm)

Family name: Thripidae
Latin name:
Many different species

LIFE CYCLE

The adults overwinter in plant debris. After mating, they insert eggs into the tender new growth and flowers of plants. The larvae go through two instars while feeding, then move into the soil for two more instars during which they do not feed but develop wings. After that, they become adults and start the cycle all over again. There can be several generations in a year.

TYPICALLY SEEN ON

Legume family crops (beans, peas), cucurbit family (cucumbers, melons, pumpkins, squash), garlic, onions, peppers

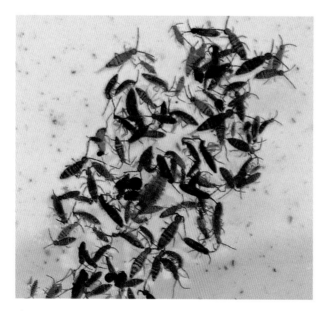

Thrips

Thrips damage on an eggplant leaf (white spots)

SIGNS OF THEIR ACTIVITY

Stunted plant growth; stippled leaves; silvery leaves often accompanied by black frass; rolled leaves; scarred fruits

NATURAL PREDATORS

Big-eyed bugs, green lacewings, minute pirate bugs, mites, parasitic wasps, predatory thrips, syrphid flies

CONTROLS

> Clean up plant debris during and at the end of the season.

> Prior to planting, consider covering the soil with a reflective silver plastic mulch, which will make it more difficult for thrips to find host plants (refer to the DIY on page 178).

> Monitor host plants regularly; this can involve holding a white piece of paper beneath a flower or leaves that you suspect thrips are on and tapping the plants to knock them onto the paper.

> Dislodge thrips with a jet of water.

> Hang blue sticky traps near host plants to capture them (refer to the DIY on page 188).

> Catch them in cardboard earwig traps (refer to the DIY on page 162).

> Control weeds in and near your garden and avoid planting near weedy or grassy areas.

> Keep plants healthy and unstressed so they can withstand damage.

> Apply beneficial nematodes, diatomaceous earth, insecticidal soap, kaolin clay, Neem, pyrethrins, or spinosad.

WHITEFLY

Whiteflies are tiny, $\frac{1}{16}$-inch-long (2 mm) moths with pale yellow bodies and four white wings. They use their piercing mouthparts to extract the sap from plants. Despite their name, they aren't flies but are related to aphids, mealybugs, and scale. In warm weather, their population really takes off, making them extremely difficult to control. Because of their tiny size, it is hard to notice whiteflies until they establish a huge colony. Many species damage vegetable crops. Just like aphids, whiteflies produce a sticky honeydew of excess liquid while feeding. This can lead to black sooty mold growing on the surface of the leaves, which interferes with photosynthesis. Whiteflies are a common problem in commercial greenhouses. The silverfleaf whitefly (*Bemisia argentifolia*) transmits tomato plant viruses, such as yellow leaf curl and tomato mottle, as well as golden mosaic virus on beans.

$\frac{1}{16}$" (2 mm)

Family name: Aleyrodidae
Latin name:
Many different species

Whiteflies on a tomato leaf

LIFE CYCLE

Whiteflies go through a simple metamorphosis. The females lay oblong eggs on the undersides of leaves while feeding. After hatching 1 week later, the nymphs go through four instars: crawlers, two feeding stages while remaining in the same location on the leaf, and one non-feeding stage. At that point, the adults emerge with wings. The entire life cycle takes 4 to 6 weeks.

TYPICALLY SEEN ON

Artichokes, beans, cabbage family crops (arugula, broccoli, Brussels sprouts, cabbage, cauliflower, kale, kohlrabi, mustard, radishes, rutabagas, turnips), cucumbers, nightshade family crops (eggplants, peppers, potatoes, tomatillos, tomatoes), peanuts, squash, sweet potatoes

Whitefly damage on a broccoli leaf

Whiteflies are tiny and can be difficult to see.

SIGNS OF THEIR ACTIVITY

Leaves turn yellow, may be stunted, or drop off the plant; mottled bean leaves; cloud of white flies when plant is disturbed; honeydew on the leaves; black sooty mold growing on the surface of the leaves; uneven ripening of tomatoes

NATURAL PREDATORS

Big-eyed bugs, damsel bugs, lacewings, ladybugs, minute pirate bugs, parasitic wasps

CONTROLS

> Always check the undersides of the leaves of plants before purchasing them from a greenhouse.

> Consider covering the soil with a reflective plastic mulch prior to planting as a means to repel whiteflies. (Refer to the DIY project on page 178.)

> Monitor plants frequently.

> Place yellow sticky traps near target crops to monitor for whitefly activity; note that the traps will only catch flying adults. (Refer to the DIY project on page 188.)

> Remove and dispose of any infested leaves; this can help natural predators manage the remaining population.

> Spray whiteflies off plants with a jet of water.

> Apply horticultural oils, insecticidal soaps, Neem, plant extracts, or pyrethrins.

WIREWORM

Click beetles are the adult form of wireworms and are so named for the "click" sound they make when flexing their prothorax (the area immediately behind the head) in order to flip themselves over. The beetles are bullet-shaped, brown or black, and up to ½ inch (1.3 cm) long. They do not cause damage to plants, but their larvae can be very destructive to plant roots and root crops. Their tunneling activities can adversely affect plant health and make root crops unappealing. The light brown wireworms grow up to ¾ inch (2 cm) long and develop hard, crosswise bands as they mature.

¾" (2 cm)

Family name: Elateridae
Latin name: Many different species

Click beetle

LIFE CYCLE

Click beetles overwinter in the soil and emerge in the spring. They lay eggs in the soil of grassy areas and usually die about 2 weeks later. Wireworms are initially white for the first year or two and eventually change to light brown as they mature. They can live at least 6 years and will overwinter in the soil, moving to different depths as the temperatures fluctuate. Once wireworms are completely mature, they will pupate within a cell in the soil, later emerging as adults. The complete life cycle for wireworms can last several years.

TYPICALLY SEEN ON

Primarily carrots, parsnips, potatoes, sweet potato tubers; also found in the roots of beans, corn, cucumbers, leeks, lettuce, melons, onions, pea plants

Wireworm larva

Wireworm damage to a potato

SIGNS OF THEIR ACTIVITY

Straight, round holes in carrots, parsnips, potatoes, and sweet potatoes; stunted or wilting plants.

NATURAL PREDATORS

Birds, fly larvae

CONTROLS

> Note that wireworms can be prevalent in new garden areas that were previously planted with sod. Monitor plants during the season for stunted growth or wilting.

> Use one of the following baiting methods to catch wireworms: 1) cut up a potato or carrot and attach each piece to a wooden pick; place them in the soil 2 to 4 inches (5 to 10 cm) deep; leave in place for one week, then dig up and destroy any wireworms caught in the vegetable chunks; 2) germinate (sprout) some pea, bean, or corn seeds; plant them 2 to 4 inches (5 to 10 cm) deep and up to 10 inches (25 cm) apart; cover each with a board so you can find the seeds later; dig up the seeds in a week and kill the wireworms that are feeding on them.

> Apply beneficial nematodes or plant extracts.

> Clean up garden debris at the end of the season.

PROFILES OF BENEFICIALS

ASSASSIN BUG (Reduviidae) (*adults only*)
Size: ¼"–1" long (1.3–2.5 cm)

PESTS IT EATS	PHYSICAL FEATURES	HOW TO ENCOURAGE
Aphids, cabbage worms, Colorado potato beetles, cucumber beetles, cutworms, earwigs, hornworms, Japanese beetles, leafhoppers, Mexican bean beetles, stink bugs	Assassin bugs have small heads, spiny legs, and a prominent beak used to pierce the bodies of their prey and inject toxin. Many species are brown or black but there are brightly colored species as well.	A healthy, pesticide-free landscape with trees, shrubs, and flowers (goldenrod, sunflowers, and wildflowers) will attract them. They often hunt on flowers.

BIG-EYED BUG (Geocoridae) (*adults and nymphs*)
Size: ⅙" long (4 mm)

PESTS IT EATS	PHYSICAL FEATURES	HOW TO ENCOURAGE
Aphids and their eggs, beet armyworms, cabbage loopers, corn earworm eggs, flea beetles, harlequin bugs, leafhoppers, Mexican bean beetles, spider mites and their eggs, thrips, whiteflies	These true bugs have stout bodies with short, wide heads and bulging eyes. Their 2 pairs of clear wings cross to form a "V" pattern. They use their piercing/sucking mouthparts to kill prey and extract their innards. Nymphs are smaller and wingless.	Create a diverse landscape to provide habitat. Plant cover crops and flowers to supplement their diet with nectar, pollen, and seeds.

DAMSEL BUG (Nabidae) (*adults and nymphs*)
Size: ¼"–1" long (1.3–2.5 cm)

PESTS IT EATS	PHYSICAL FEATURES	HOW TO ENCOURAGE
Aphids, beet armyworms, cabbage worms, Colorado potato beetle eggs and larvae, corn earworms, cutworms, various insect eggs, leafhoppers, whiteflies	This generalist predator has a long, slender brown body, 2 pairs of wings, piercing/sucking mouthparts, and spiny front legs for grasping prey. They also have bulging eyes.	They overwinter in leaf litter so keep leaves under shrubs. They also find grassy areas, cover crops, and low-growing shrubs or flowers to ambush their prey attractive.

DRAGONFLY (Suborders: Anisoptera; Zygoptera) (*adults and nymphs*); **DAMSELFLY** (*adults only*)
Size: ¾"–3" long (dragonflies, 2–7.5 cm), up to 1½" long (damselflies, 3.8 cm)

PESTS IT EATS	PHYSICAL FEATURES	HOW TO ENCOURAGE
Beetles, mosquitoes, moths	These aerial acrobats mostly catch prey on the wing. Both have 2 pairs of transparent wings, long abdomens, and huge eyes. Dragonflies hold their wings open at all times, fly forward, backward, and hover; damselflies fold their wings when at rest. Both have lengthy aquatic nymph stages.	Ponds and other water features in full to part sun attract both. Provide plants with upright stems as hunting perches for them. Dragonfly nymphs eat mosquito larvae in the water.

GROUND BEETLE (Carabidae) *(adults and larvae)*
Size: ⅛"–1 ½" long (0.3–3.8 cm)

PESTS IT EATS	PHYSICAL FEATURES	HOW TO ENCOURAGE
Aphids, carrot rust flies, Colorado potato beetles, corn earworms, cutworms, diamondback moths, root maggots, slugs and snails, squash vine borers	These nocturnal beetles have black bodies with some iridescent coloration, threadlike antennae, and large mandibles. The wing covers have ridges.	Provide rocks or logs under which beetles can hide. Avoid tilling the soil. Add a "beetle bank" to your landscape by creating a raised mound and planting it with native bunch grasses.

HOVERFLY (Syrphidae) *(a.k.a., flower fly or syrphid fly)* *(larvae only)*
Size: ¼"–½" long (0.6–1.3 cm)

PESTS IT EATS	PHYSICAL FEATURES	HOW TO ENCOURAGE
Aphids, cabbage loopers, cabbage worms, leafhoppers, thrips	Often mistaken for bees or wasps, these flies have black or brown and yellow stripes and only 1 pair of wings. They hover above flowers to feed on nectar and pollen. Their larvae, which are green or cream-colored maggots with tapered heads, are the predators.	Plant a variety of flowering annuals, perennials, and herbs that bloom all season long. Avoid tilling the soil.

LACEWING (Green: Chrysopidae; brown: Hemerobiidae) *(adults and larvae)*
Size: green: ½"–1" long (1.3–2.5 cm); brown: ⅓"–½" long (0.8–1.3 cm)

PESTS IT EATS	PHYSICAL FEATURES	HOW TO ENCOURAGE
Aphids, asparagus beetle larvae, cabbage worms, Colorado potato beetle larvae, corn earworms, harlequin bugs, hornworms, various insect eggs, leafhoppers, thrips, whiteflies	Lacewings have slender bodies and large, transparent wings held over their bodies when at rest. Green lacewings have green bodies and wings; brown lacewings have light brown bodies. Both have alligator-like larvae. Green lacewings lay eggs on stalks.	Plant flowers that bloom all season to provide nectar and pollen; also cover crops, such as crimson clover.

LADYBUG (Coccinellidae) *(adults and larvae)*
Size: ¼" long (6 mm)

PESTS IT EATS	PHYSICAL FEATURES	HOW TO ENCOURAGE
Aphids, asparagus beetles, Colorado potato beetle larvae, corn earworms, harlequin bugs, hornworms, various insect eggs, leafhoppers, Mexican bean beetle larvae, whiteflies	Adults have round bodies, most with brightly colored wing covers and black spots, although there are a surprising number of variations. The red- or orange-and-black larvae look like small alligators (see page 17). Eggs are yellow or orange.	Keep leaf litter in place under shrubs and landscape beds to provide overwintering habitat. Plant a wide variety of flowers along with flowering herbs, such as cilantro, dill, and thyme.

LONG-LEGGED FLY (Dolichopodidae) *(adults and possibly larvae)*
Size: ¼" long (6 mm)

PESTS IT EATS	PHYSICAL FEATURES	HOW TO ENCOURAGE
Adults prey on aphids, spider mites, thrips. Larvae are believed to be predaceous as well.	These beautiful flies have metallic green or copper bodies, prominent red eyes, and elongated legs. Wings often have black markings.	Wetlands and meadows attract these flies. Provide nectar sources with a diverse range of flowers.

MINUTE PIRATE BUG (Anthocoridae) *(adults and nymphs)*
Size: ⅛" long (3 mm)

PESTS IT EATS	PHYSICAL FEATURES	HOW TO ENCOURAGE
Aphids, beet armyworms, corn earworms, harlequin bugs, various insect eggs, leafhoppers, Mexican bean beetle larvae, thrips, whiteflies	Adults have oval black bodies with white markings on their backs. The nymphs' orange bodies are teardrop-shaped and they have red eyes.	Adults require the pollen and sap from spring-flowering plants when they first emerge in spring. Plant cover crops, such as buckwheat and crimson clover, and early blooming perennials (i.e., basket of gold, candytuft, coreopsis, daisies, and yarrow).

PARASITIC WASP (Hymenoptera) (*adults and larvae*)
Size: 1/32"–1/2" long (0.08–1.3 cm)

PESTS IT EATS	PHYSICAL FEATURES	HOW TO ENCOURAGE
Aphids, asparagus beetles, cabbage loopers and worms, carrot rust flies, Colorado potato beetles, corn earworms, cucumber beetles, cutworms, diamondback moths, flea beetles, harlequin bug eggs, hornworms, Japanese beetles, leafminers, Mexican bean beetles, root maggots, squash vine borers, stink bugs, thrips, whiteflies	There are many species, with the most common being Braconid (pictured), Chalcid, Ichneumon, and Trichogramma. They have slender bodies with narrow waists, threadlike antennae, and are black, red, or metallic. Some females have an ovipositor for laying eggs inside their prey; it is not a stinger. When the eggs hatch, the larvae devour their host.	Provide pollen and nectar sources with flowers in the aster, carrot, and legume families.

PRAYING MANTIS (Mantidae) (*adults and nymphs*)
Size: Up to 5" long (12.7 cm)

PESTS IT EATS	PHYSICAL FEATURES	HOW TO ENCOURAGE
Aphids, asparagus beetles, Colorado potato beetles, earwigs, grasshoppers, leafhoppers, Mexican bean beetles, squash bugs, stink bugs	Praying mantids have long, slender bodies and spiny front legs for capturing prey. They are green or brown, and have triangular, swiveling heads with large eyes. They make their brown egg cases from a hardened foam and attach them to sticks, branches, or structures. Nymphs hatch as miniature versions of adults.	Watch for egg cases on branches and twigs. Leave them in place or clip the stem to move them to a sheltered place in the garden.

PREDATORY STINK BUG (Pentatomidae) (*adults and nymphs*)
Size: Up to ½" long (1.3 cm)

PESTS IT EATS	PHYSICAL FEATURES	HOW TO ENCOURAGE
Many kinds of caterpillars, Colorado potato beetle eggs and larvae, Japanese beetles, Mexican bean beetles, plant-eating stink bugs	Predatory stink bugs have the distinctive shield-shaped bodies and unpleasant odor of all stink bug species. Adults vary from pale brown to black with yellow or red markings, and have short, wide "beaks." Nymphs are red and black, changing to yellow, tan, or black as they mature. Eggs are barrel-shaped.	Adults overwinter in leaf litter. Provide a diverse landscape of trees, shrubs, and flowers to provide shelter.

ROBBER FLY (Asilidae) (*adults and larvae*)
Size: 1"–1½" long (2.5–3.8 cm)

PESTS IT EATS	PHYSICAL FEATURES	HOW TO ENCOURAGE
Aphids, Colorado potato beetles, grasshoppers, Japanese beetles, leafhoppers, Mexican bean beetles	Robber fly species vary in appearance: Some have long, slender bodies while others mimic the appearance of bumblebees. All have 2 wings and long, bristly legs. Their bodies are brown, gray, or black. The females have long ovipositors for laying eggs.	They don't require pollen or nectar; provide good habitat with a diverse landscape of trees, shrubs, and perennial and annual flowers.

ROVE BEETLE (Staphylinidae) *(adults and larvae)*
Size: ¼"–1" long (0.6–2.5 cm)

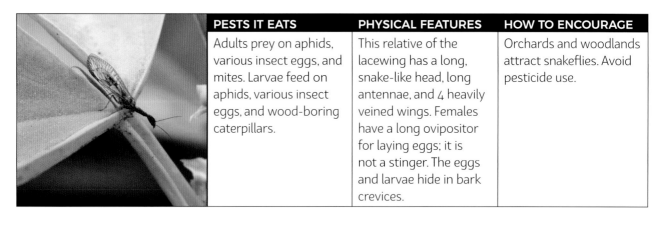

PESTS IT EATS	PHYSICAL FEATURES	HOW TO ENCOURAGE
Carrot rust flies, caterpillars, root maggots, slugs and snails, spider mites	Rove beetles have slender, elongated black or brown bodies with short wing covers that leave their segmented abdomen exposed. Those segments and the tapered end make them look similar to earwigs, but they do not have pincers.	They prefer moist locations under rocks, logs, brush piles, and leaf litter, so leave habitat for them within your landscape. Flowers attract rove beetles.

SNAKEFLY (Raphidiidae) *(adults and larvae)*
Size: ½"–1" long (1.3–2.5 cm)

PESTS IT EATS	PHYSICAL FEATURES	HOW TO ENCOURAGE
Adults prey on aphids, various insect eggs, and mites. Larvae feed on aphids, various insect eggs, and wood-boring caterpillars.	This relative of the lacewing has a long, snake-like head, long antennae, and 4 heavily veined wings. Females have a long ovipositor for laying eggs; it is not a stinger. The eggs and larvae hide in bark crevices.	Orchards and woodlands attract snakeflies. Avoid pesticide use.

SOLDIER BEETLE (Cantharidae) *(a.k.a., Leatherwings) (adults and larvae)*
Size: ½" long (1.3 cm)

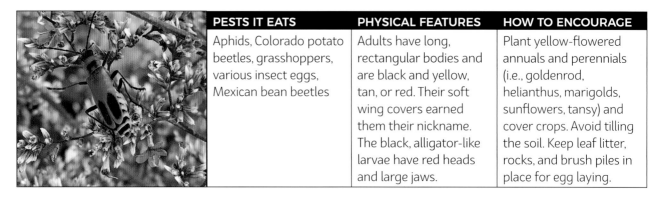

PESTS IT EATS	PHYSICAL FEATURES	HOW TO ENCOURAGE
Aphids, Colorado potato beetles, grasshoppers, various insect eggs, Mexican bean beetles	Adults have long, rectangular bodies and are black and yellow, tan, or red. Their soft wing covers earned them their nickname. The black, alligator-like larvae have red heads and large jaws.	Plant yellow-flowered annuals and perennials (i.e., goldenrod, helianthus, marigolds, sunflowers, tansy) and cover crops. Avoid tilling the soil. Keep leaf litter, rocks, and brush piles in place for egg laying.

SPIDER (Order Araneae) (*adults and juveniles*)
Size: ¼"–1½" long (0.6–3.8 cm)

PESTS IT EATS	PHYSICAL FEATURES	HOW TO ENCOURAGE
Aphids, asparagus beetles, cabbage worms, Colorado potato beetles, corn earworms, cutworms, diamondback moths, harlequin bugs, leafhoppers, leafminers, pillbugs and sowbugs, squash bugs, stink bugs	Spiders have 2 body parts—the cephalothorax and abdomen—8 legs, and 4 pairs of eyes. Colors vary widely. Baby spiders eat as voraciously as their parents.	They overwinter in brush piles and clumps of grass, and will remain in a healthy environment with plenty of prey. Plant cover crops and flowering plants.

TACHINID FLY (Tachinidae) (*larvae only*)
Size: Up to ¾" long (2 cm)

PESTS IT EATS	PHYSICAL FEATURES	HOW TO ENCOURAGE
Beet armyworms, cabbage loopers, cabbage worms, Colorado potato beetle larvae, corn earworms, cucumber beetle larvae, cutworms, diamondback moth larvae, earwigs, Japanese beetle larvae, Mexican bean beetle larvae, squash bug larvae	Tachinid flies come in gray, brown, yellow, red, or metallic colors. They have rounded bodies and hairy abdomens.	Adults require nectar and pollen so plant flowers that bloom all season, including those of the carrot, aster, and milkweed families. Keep leaf litter for larvae to pupate in.

There's a reason this head of broccoli is pest free! Find out why on page 145.

3 ORGANIC PEST MANAGEMENT PRODUCTS AND DIY PEST CONTROLS

In the previous chapter, I identified specific vegetable garden pests and the many options you have for controlling them. In this chapter, we'll get down to the nuts and bolts of the control methods and how to implement them. In the first half are details on easy-to-find organic products available to gardeners. The second half is filled with economical projects you can make to keep away pests or lure them into a trap and some very cool items that will add greatly to your enjoyment of gardening.

Remember our discussion of Integrated Pest Management (IPM) (page 14)? Before you take action, know exactly which garden bug you have, determine whether it's a pest, and confirm that the product or method you intend to use will address the problem. Always choose the most environmentally friendly method first. Using control products—even when they're labeled as organic—should be your last resort. This is especially important since we're talking about edible crops. Be sure to read and follow the label directions: Manufacturers are legally required to list the least amount of the product necessary in order to be effective. Avoid getting into the mindset of thinking, "if 2 tablespoons of this product in a gallon of water is effective, 4 tablespoons must be even better!" This waste of the product (and your money) increases its toxicity and the likelihood it will end up in groundwater.

I strongly believe that if you follow the tried-and-true cultural practices listed in chapter one, your plants will thrive and be better able to survive damage from pests.

The other important strategy is to use simple techniques to control pests that cause trouble. My two favorite methods are removing pests by hand and excluding them with a barrier. Each of these works well and without a detrimental effect on other species or the environment. You'll learn the specifics of using row covers on page 145, but let's talk about hand-picking for a moment.

This method works great when the pests causing damage are present in manageable quantities. Fill a small pail or container with water and add a couple of squirts of dish soap. It's easy to knock pests such as beetles, caterpillars, and worms into the container and the soap coats their bodies so they can't fly or crawl out. If you're squeamish and the pest is rather disgusting—let's say it's a massive tomato hornworm—wear gloves and either pull it off the plant or snip off the branch it's on and dispose of it.

Another hands-on control is to just squish the problem. Remember that an aphid infestation begins with a female and her cluster of young. The whole group can be killed by smashing them with your thumb. Your hose is a useful tool as well: A small blast of water on aphids or whiteflies will knock them off a plant. Early suppression of these pests prevents them from becoming a bigger problem later.

There are many other straightforward strategies you can employ.

> **Change the timing of your plantings.** If your growing season is long enough to allow you some flexibility, consider planting crops earlier or later than is customary to avoid a pest's egg-laying season. Within the pest profiles in chapter two, I pointed out the pests to which this approach applies. If you have a short growing season, this strategy probably won't be an option.

> **Start seedlings indoors.** This will get them off to the best start possible before you transplant them out in the garden and make them better able to tolerate early pest damage. You might be surprised to hear I do this with my beans, corn, melons, peas, and squash, but it's well worth the effort.

> **Plant "trap crops."** This method involves luring a dreaded pest away from a prized crop by attracting it to a temporary planting. Dispose of the plants, pests, and (hopefully) their eggs and/or larvae before they move on. Here are two ways to accomplish this: 1) Plant a crop that is attractive to your problem pest in another area of your garden, away the crop you want to protect. After they congregate on the trap crop, dispose of the plants, pests and all. 2) Start a few seedlings of the targeted crop earlier than usual, allow the pests to feed on them, then eliminate the plants and pests. Plant the trap crops at least 2 weeks ahead of time. Both approaches work quite well for controlling flea beetles, Mexican bean beetles, squash bugs, and squash vine borers. Consider planting radishes to attract flea beetles, extra bean plants to lure in bean beetles, or spare squash plants to draw in those two horrid squash pests.

> **Skip a year or two of growing certain crops.** Some pests are particularly bad news in the garden: Colorado potato beetles, squash bugs, squash vine borers, and root maggots immediately come to mind. If any of these have become an annual problem, why not skip planting the crops they target for a year or two? This disrupts the pest's life cycle and, while it might sound drastic, can resolve the annual problem of having to deal with the damage they cause. Even though you might miss those crops, perhaps you can pick them up at local farmers' markets and take the opportunity to plant another vegetable that you've always wanted to try!

> **Don't forget about interplanting.** Adding aromatic plants within your vegetable garden may repel or confuse pests, expand the diversity of what you're growing, and look attractive, too. Examples include catnip, lavender, marigolds, members of the onion family, and sage. I plant onion sets (bulbs) in rows as borders to define different crops within the same bed.

> **Don't bring trouble home.** When purchasing vegetable seedlings from a greenhouse, nursery, or plant sale, always inspect them thoroughly. So many gardeners have told me how they never had a problem with a certain pest until it hitched a ride to their garden on a plant they bought. This has prompted me to grow all of my vegetable crops from seed, just to be on the safe side.

In addition to considering the above tactics, once you've read through the descriptions of the organic products in the coming pages, you will have all of the details you need to make wise pest control choices.

I want to underscore a very important point: When I talk about commercial organic control products, please keep in mind that many of them will kill both beneficials and pests. The word "organic" on the labels means they are not harmful to our environment. Products that are harmful to pollinators should not be used around flowering plants. One other alert: Always check the label to determine when it is safe to harvest and consume treated vegetables.

You might wonder why I even included the commercial products in this book but the fact is, they are commonly used and are a viable alternative to synthetically produced chemical pesticides. I feel it's important to give you the details about each so you can make an informed choice.

After reading the descriptions of the organic products, check out the DIY projects, which are designed to catch, repel, or exclude pests. All of them are quick and easy to make and certain to give you the satisfaction that you're dealing with the damaging pests in your garden. Sure, you can purchase most of these items from commercial sources, but isn't it more rewarding—and economical—to use your creativity and make your own?

HERE'S WHAT YOU'LL FIND:

> If snails and slugs are an issue in your garden, make traps or barriers. See pages 148 and 154.

> Are cutworms wiping out tender seedlings? Find out how to protect the plants until they can fend for themselves. See page 160.

> Use everyday items to create traps that earwigs won't be able to resist. See page 162.

> Lure those bothersome cucumber beetles to their deaths in traps containing botanical oils. See page 156.

> Discover how well sticky traps work to control many problematic pests. See page 188.

> Learn how to use reflective plastic mulch to confuse pests. See page 178.

> Erect a simple barrier to keep carrot rust flies away from your crops. See page 152.

If you're looking for some "meaty" projects that will enhance your gardening experience, check out these cool projects:

> Attract pollinators and other beneficials by assembling your very own insect hotel (warning: making them is addicting!). See page 165.

> Build a covered raised bed designed to effortlessly keep away troublesome pests. See page 170.

> Make a pipe bender to create your own sturdy metal row cover hoops or create very inexpensive yet durable hoops from plastic sprinkler pipe. See page 182.

Each DIY project contains clear instructions and photographs to illustrate the steps. It is my hope you'll feel a sense of satisfaction when the end product catches pests, keeps them away, or brings in beneficials.

BACILLUS THURINGIENSIS (BT)

Bacillus thuringiensis, or *Bt* as it's commonly called, is a long name for an effective organic insecticide. First discovered in Japan in 1901, *Bt* is a naturally occurring soil bacterium that gardeners and farmers apply to the foliage of host plants. Once ingested by certain pests, it paralyzes their gut, causing them to stop feeding within minutes. They starve to death over the course of a few days.

Gardeners have access to three *Bt* strains: *kurstaki*, *galleriae*, and *israelensis*.

Bt kurstaki (*Btk*) controls the caterpillar (larval) stage of specific moths and butterflies. It is effective against beet armyworms, cabbage loopers, cabbage worms, corn earworms, cutworms, diamondback moth caterpillars, tomato fruitworms, and tomato hornworms. *Bt* does not affect the adult form of these insects or their eggs.

Since it is more effective on young caterpillars, watch their host plants and apply *Btk* as soon as you notice their presence. Most damaging larvae are initially found on the undersides of the leaves so be sure to apply *Btk* to all leaf surfaces. It is available in a liquid concentrate that you mix with water, as a pre-mixed spray, or as a dust. You may apply it on the day of harvest, although you should always wash produce before consuming it. *Btk* won't harm humans, birds, earthworms, mammals, or other insects. It doesn't differentiate between "good" caterpillars and bad ones, however, so use wisely.

No matter which form of *Btk* you buy, follow the label directions to the letter. Only apply the product when you have a problem with the abovementioned insect pests, rather than "just in case." Research has shown that some caterpillars—such as diamondback moth larvae—are developing resistance to *Btk* so it is important to use it judiciously and alternate it with other control methods.

Bt galleriae (*Btg*) helps control both the adult and larval form of Japanese beetles. Apply it to turfgrass when the grubs hatch and again when the beetles first emerge. Refer to the label for the timing of applications. Research studies have found that *Btg* is highly toxic to the caterpillars of monarch butterflies, so do not apply anywhere near their milkweed host plants.

Bt israelensis (*Bti*) effectively controls the larvae of mosquitoes, fungus gnats, and black flies. Because fungus gnats are quite troublesome in the soil of both greenhouse plants and houseplants, this strain might be of interest to you.

Even though these strains of *Bt* are considered safe for use around pollinators, I prefer to err on the side of caution by avoiding spraying near flowers, and I apply *Bt* late in the day when bees aren't active. Another good reason for spraying late in the day is the slower evaporation rate: This ensures it will remain on the foliage longer. Refer to the label recommendations for reapplication intervals.

All three strains are sold in garden centers and online under several brand names. Look for suggested online sources starting on page 192.

BENEFICIAL NEMATODES

Nematodes are soil-dwelling, microscopic roundworms. Just as there are good bugs and bad bugs, there are beneficial and damaging nematodes. Gardeners can purchase the beneficial ones from garden centers and online suppliers to control many troublesome pests.

As predators, they locate hosts within the soil by detecting the pest's respiration. The nematodes enter through openings in the host's body (such as the mouth, breathing holes, or anus) and infect that host with the bacteria they carry. This kills the host within 48 hours. Once the host dies, larvae that are in the third stage leave the body in search of another host.

Beneficial nematodes are a good alternative for dealing with pests that have developed resistance to other controls. They are safe to use around humans, animals, birds, pollinators, other beneficials, and earthworms. The nematodes only parasitize the larval or grub stage of insects that are in the soil.

The beneficial nematode species most commonly used to target vegetable garden pests are *Heterorhabditis bacteriophora*, *Steinernema carpocapsae*, *S. feltiae*, and *S. riobrave*. Refer to the table for a list of the pests each species controls. Always consult with a supplier to make the best choice for your pest problem.

Apply nematodes during the spring or fall when the damaging larvae or grubs are in the soil. To assist with the timing, refer to the pest profiles in chapter two.

When you purchase beneficial nematodes, follow these important guidelines:

> Store them under ideal conditions listed on the accompanying literature.
> Nematodes require a moist environment in order to move through the soil as they search for hosts. Moisten the soil prior to the treatment.
> Use the nematodes as soon as possible. Choose an overcast day when the humidity is high and the soil temperature is above 42°F (5.6° C). The best times of day are in early morning or late in the day to ensure the nematodes won't dry out.
> Mix the nematodes with water in a clean sprayer or watering can; apply them to the soil within two hours.
> Keep the soil moist by watering the area regularly over the following 2 weeks.

Heterorhabditis bacteriophora	Asparagus beetles, carrot weevils, Colorado potato beetles, cucumber beetles, flea beetles, Japanese beetles, leafminers, squash vine borer*, wireworms
Steinernema carpocapsae	Beet armyworms, carrot weevils, caterpillars, corn earworms, cutworms, earwigs, fly larvae, leafminers, pillbugs, sowbugs, wireworms
Steinernema feltiae	Beet armyworms, cabbage maggots, carrot weevils, corn earworms, cucumber beetles, cutworms, leafminers, onion maggots, pillbugs, root maggots, thrips, wireworms
Steinernema riobrave	Carrot weevils, cutworms, Japanese beetles

*Note

Inject beneficial nematodes directly into the main stem of cucurbit family crops to control squash vine borers.

DIATOMACEOUS EARTH

Diatomaceous earth may resemble flour, but it holds a secret gardeners have employed for more than 100 years. Its main ingredient is the pulverized, fossilized remains of diatoms, which are tiny algae found all over the world in bodies of water and the soil. Diatoms are the only life forms on this planet with cell walls made of silica, and it's the microscopic sharp edges of this mineral that make it an effective barrier to pests.

It is safe for humans to handle, although we should take precautions to avoid inhaling it. When any small bug comes into contact with diatomaceous earth, those sharp edges cut into its skin, causing the creature to dehydrate and die. The product will also break down a bug's protective coating (cuticle), ending with the same result. Diatomaceous earth is safe to use around pets, which is always an important consideration.

The label on the packaging states it controls crawling insects, such as ants and earwigs, but there are a lot of applications for its use in the garden. It can prevent or stop damage from aphids, beetles, caterpillars, cutworms, leafhoppers, pillbugs, slugs, snails, and thrips. Some folks use it indoors to control fleas, bedbugs, and roaches.

Here are the two ways I use it: Once I plant a seedling, I invert an empty can over it and sprinkle a ring of diatomaceous earth around it to create a barrier. Then I remove the can, repeat the process with the next seedling, and so on. My alternate method is to sprinkle it immediately around the stem of each plant, especially when I'm trying to protect the tender stems of young seedlings from cutworms.

Diatomaceous earth doesn't distinguish between good and bad bugs. It is safe to sprinkle diatomaceous earth on the plants themselves, but avoid getting it on flowers to protect pollinators. To apply it to leaves and stems, lightly mist them with water first so the product

Create a barrier around plants that pests won't cross.

will adhere. When using diatomaceous earth on the soil surface, I find that once it rains or gets particularly wet, it is less effective and has to be reapplied.

Garden centers sell the food grade version of diatomaceous earth, which is the safest grade to use and handle. Be sure to store it in a dry location.

HORTICULTURAL OIL

Gardeners use horticultural oils made from mineral or vegetable oil to control some types of insects and mites. When thoroughly applied to plant foliage where pests or eggs are present, the oil blocks the breathing holes (spiracles) along the pest's abdomen and smothers the eggs. It is particularly effective on soft-bodied pests and the egg stage of aphids, spider mites, and whiteflies. The oil also kills the early stages of some nymphs and caterpillars, provided they are well-coated.

These products are labeled for use on vegetable plants as well as ornamental trees and shrubs, roses, tree fruits, berries, and nut trees. You can apply horticultural oils year-round within the landscape and gardeners frequently use them on dormant trees and shrubs to smother the eggs of aphids, spider mites, and scale. In the vegetable garden, apply oil to a plant's leaves and stems during the growing season. Some sprays are safe to use up to the day before harvest, but always consult the label first for this information.

As with all products, read the label thoroughly and take precautions to avoid contact with your skin and eyes. Wear long sleeves and pants, gloves, and a pair of safety goggles. Stay away from the sprayed area until the oil has dried thoroughly. Keep pets away as well. Because horticultural oils are toxic to fish, do not spray near bodies of water, such as ponds, creeks, or lakes.

Use oil only when damaging pests or their eggs are present and always use the recommended application rate on the label. Avoid spraying flowers and time the application for late in the day or in early morning before pollinators are active. Be aware that horticultural oil might damage young seedlings and should never be used on diseased plants.

Apply the oil during calm weather conditions, to avoid spray drift, and when the weather forecast indicates it won't rain for at least 24 hours. Never use horticultural

oil during drought conditions or when temperatures are above 85°F (29°C). Why? The oil affects a plant's respiration—where it converts sugars made through photosynthesis into energy—and transpiration, which is the movement of water throughout its tissue.

Horticultural oils are available as a pre-mixed spray or a concentrate that you mix with water. During the application, shake the mixture to prevent the water and oil from separating, which would reduce its effectiveness.

Prior to spraying, irrigate the plants so they are sufficiently hydrated. It's a good idea to test-spray a small area of the plant first, to make sure it will tolerate the oil without incurring any damage. If the plant is fine, spray all surfaces of its leaves.

Don't let sprinklers hit your treated plants since the water will reduce the oil's effectiveness. Refer to the label to learn when it is safe to reapply the oil, if necessary.

INSECTICIDAL SOAP

Insecticidal soap, made from potassium salts of fatty acids, is a useful organic tool for controlling many damaging vegetable garden pests. The fatty acids first break down the affected bug's protective shell (cuticle), then penetrate the cell membranes, causing dehydration.

The soap only works on contact with the pest. Once it has dried, it is no longer effective. Insecticidal soap is most effective on soft-bodied pests, such as aphids, spider mites, and whiteflies, but also controls earwigs, grasshoppers, leafhoppers, lygus bugs, squash vine borers, stink bugs, and thrips. Pests rarely develop resistance to the spray.

While it might be tempting to make your own insecticidal soap using household detergents, I would advise against it. Dish soaps and laundry detergents are too harsh and can burn plant foliage. Commercially available insecticidal soaps are specifically designed for use in the garden and are safe around humans, birds, and mammals.

However, they are toxic to pollinators and other beneficials while the spray is wet. To avoid harming these beneficials, apply insecticidal soaps in the evening or early morning before they become active. Do not spray open flowers or near water sources.

When you have a pest problem in your vegetable garden, be sure to properly identify it first, then read the labels of insecticidal soaps to locate one that will control it. These products are available in concentrates to mix with water and as ready-to-use, pre-mixed sprays. They are easy to find at garden centers, hardware stores, and home centers.

To be on the safe side, test-spray a small area of the plant 48 hours before you intend to fully treat it. This way, you'll know if the oil will burn the leaves. Beans, cucumbers, and peas are more susceptible to damage from the spray. If you notice wilting, gently hose down

the test plant. It is important to avoid spraying young seedlings.

Choose a calm day when plants are dry but well-hydrated. Do not apply insecticidal soaps during drought conditions, when the temperature is above 90°F (32°C), or if there is high humidity.

Follow the mixing and application directions on the label. Remember that using more product than the instructions call for will not be more effective and could damage the plants. Always start with a clean sprayer to mix the concentrated solution with water.

When spraying plants, coat the upper and lower surfaces of the plant's damaged leaves and any pests you see. Remember that it is only effective while it's wet.

You might need to reapply insecticidal soaps to control a problem pest. Refer to the label directions for information on how soon to repeat the application.

KAOLIN CLAY

Kaolin clay is a naturally occurring mineral that is ground into a powder, mixed with water, and applied to the leaves and fruits of vegetable crops. It creates a protective coating on plant foliage that both confuses the pest ("hey, this isn't an eggplant!") and irritates their skin. This inhibits feeding on the leaves or fruits, as well as deterring egg-laying.

Also known as china clay, kaolin clay is used in the manufacturing of china and porcelain dishes. The name kaolin comes from the Chinese word *gaoling* ("high hill"), which refers to the location where the clay was first discovered and mined in the 1700s. Do not purchase the pottery grade of kaolin clay for pest control because the larger particles will damage plants and clog sprayers. There is a much more refined version for use in the garden that has the copyrighted name Surround.

Kaolin clay controls asparagus beetles, Colorado potato beetles, cucumber beetles, flea beetles, grasshoppers, Japanese beetles, leafhoppers, lygus bugs, Mexican bean beetles, spider mites, squash bugs, squash vine borers, stink bugs, and thrips. It is also useful to prevent sunburn and heat stress on some fruiting vegetable crops, such as eggplants, melons, peppers, squash, and tomatoes. Organic orchardists use kaolin clay as a means for controlling apple codling moths, apple maggot flies, and other tree fruit pests.

It is safe for humans, animals, and pollinators and other beneficials, and can be used up to the day of harvest. For the clay to be effective, spray plants before problems occur. Avoid applying it near water sources.

Since kaolin clay has a fine, flourlike consistency, it's a good idea to wear a dust mask and goggles. The application rate is 3 cups (924 g) of the clay to 1 gallon (3.8 L) of water. Mix them together in a bucket first to

create a slurry, then pour it into a clean sprayer. While spraying, continually agitate the mixture to prevent the clay from separating from the water.

For pest control, coat the upper and lower surfaces of the leaves and all areas of the plant. If you're protecting plants against sunburn, just coat the tops of the leaves and fruit, if desired. When you are finished, clean the sprayer and especially the nozzle since the clay tends to clog them.

Your plants won't look very pretty after the spray routine but kaolin clay is quite effective. Avoid overhead watering as this will wash the coating off the plants. Reapply after heavy rains or when the plants have new leafy growth. It is safe to reapply this clay mixture as needed. At harvest time, be sure to thoroughly wash produce before consuming it.

You can find kaolin clay at garden centers, farm supply stores, and from online sources. Refer to the resources section beginning on page 190.

NEEM

Neem is a botanical insecticide that controls a wide range of damaging pests. It is extracted from the Neem tree, which grows in Southeast Asia and India. The active ingredient, azadirachtin, affects insects and nematodes and disrupts their feeding activities. It affects the growth hormones that regulate their ability to molt and mature. Because of this, Neem is most effective on the earlier life cycles of insects. In addition, it suffocates soft-bodied pests, such as aphids, spider mites, and thrips, by blocking their breathing holes (spiracles). Neem is also labeled as a fungicide to treat problems such as powdery mildew.

The interesting thing about Neem is that plants sprayed with it take up the Neem extracts much as they would a systemic insecticide. This increases its effectiveness against insects that are particularly challenging to control. For example, leaf miner larvae tunnel between the layers of cells in leaves, which makes it difficult to impact them by spraying the surface of the plants.

While this product is safe for humans, animals, and birds, it is toxic to bees and other beneficials. Neem is also harmful to fish and other aquatic organisms so avoid spraying near bodies of water.

Just as with all organic insecticides, use them as a last resort; try other organic control methods first. It's also a good idea to alternate the use of neem with other methods to decrease the chance of insects developing a resistance to it.

Neem is easy to locate at garden centers, home centers, and online, and is available as a concentrate or as a ready-to-use spray. Look for product labels listing "clarified hydrophobic extract of Neem oil" as the active ingredient. When considering the use of Neem, always identify the insect problem first; then check the label to determine if it is effective in controlling that pest.

Always follow the label directions and use the exact amount recommended. Wear long sleeves and pants, gloves, and safety goggles to protect your skin and eyes. Do not apply Neem to wilting plants—water them ahead of time—or to young seedlings. Test-spray small areas of a plant first to determine if Neem will damage its foliage.

Since Neem is harmful to bees and other pollinators, avoid spraying blossoms, and apply it in early morning or late in the day when these insects are inactive.

It is safe to reapply Neem every 7 to 14 days, as needed.

Pests Controlled by Neem

Neem controls aphids, asparagus beetles, beet armyworms, cabbage loopers, cabbage worms, Colorado potato beetles, cucumber beetles, cutworms, diamondback moths, flea beetles, harlequin bugs, leafhoppers, leafminers, lygus bugs, Mexican bean beetles, squash bugs, thrips, tomato fruitworms, tomato hornworms, and whiteflies.

PLANT EXTRACTS

The perfume and food industries have long used extracts from aromatic plants. Plant oils and juices that contain defensive chemical compounds make quite effective organic pest control products, and gardeners use botanical oils to repel or smother and kill pests or lure them into traps. The juices from plants with natural repellent qualities can also be used to confuse or irritate pests in order to protect crops.

Because plant-based products quickly break down from exposure to sunlight and moisture, they have a low environmental impact. These products are also safer to use around humans and other animals, including birds and fish, than synthetic pesticides.

Botanical oils—such as cinnamon, clove, lemongrass, mint, and rosemary—will control aphids, cabbage worms, Colorado potato beetles, cutworms, earwigs, Mexican bean beetles, spider mites, whiteflies, and wireworms. Many interfere with feeding, egg-laying, and growth regulation. Cinnamon oil attracts Japanese beetles, leafminers, and thrips into sticky traps. Clove oil, which contains the compound eugenol, is a useful contact insecticide, and it also attracts cucumber beetles (refer to the DIY cucumber beetle trap on page 156).

Citrus oil, or limonene, works as a repellent and a contact insecticide that suffocates pests. It controls aphids, asparagus beetles, cabbage worms, Colorado potato beetles, earwigs, flea beetles, Mexican bean beetles, spider mites, and whiteflies.

As you might guess, garlic sprays are used as repellents and have long been employed to keep pests away from plants in the garden. You can find commercial sprays at garden centers and online. Garlic sprays repel pests for 1 to 2 weeks and control aphids, caterpillars, flea beetles, grasshoppers, leafhoppers, leafminers, slugs, snails, spider mites, and whiteflies.

Use hot pepper spray to repel vegetable garden pests; it acts as an irritant. The active ingredient, capsaicin,

is the chemical compound found in chili peppers that makes them so hot. When used for pest control, it breaks down cell membranes of bugs and affects their nervous systems. While it's safe for use around mammals and birds, pepper spray will kill bees. To protect them, apply it late in the evening or in the early morning, when they are inactive. Hot pepper sprays control aphids, beet armyworms, cabbage loopers, leafhoppers, spider mites, and whiteflies. Purchase commercially prepared sprays. Applications last up to 3 weeks. You will definitely want to wear gloves and goggles to protect yourself from the spray.

When using any of the above plant extracts, it's important to read the label directions, handle them properly, and dress appropriately. To avoid drift, do not spray on windy days. Always test-spray part of the plant that needs protection to determine if the spray will burn the foliage. You may need to frequently reapply these products due to their highly volatile nature.

To locate product sources, refer to the Resources section beginning on page 190.

PYRETHRINS

Even though pyrethrins are an organic insecticide, only use them as a last resort. I have included this control since you might hear about it or receive a recommendation to use it for vegetable garden pests.

Pyrethrins are derived from a chrysanthemum flower called *Chrysanthemum cinerariifolium*, commonly known as Dalmatian pellitory. Pyrethrum is a naturally occurring substance from which pyrethrins are extracted. This contact insecticide affects the nervous system of insects and other bugs, causing paralysis. While most die, some recover if they had less exposure.

When you look at the list of pests controlled by pyrethrins, you might think it's the answer to your gardening woes, but there are valid reasons why their use is controversial. The active ingredient in this broad-spectrum pesticide kills both pests and beneficials, and it is highly toxic to honeybees and aquatic life. Note that pyrethroids—synthetic compounds manufactured to mimic pyrethrins—are much more toxic to mammals and bugs and are not labeled for use in organic gardens.

You can find products containing pyrethrins at garden centers and online in concentrates, ready-to-use sprays, or dusts. Since dusts persist longer on plants than sprays, it's a good idea to select a spray in order to minimize the risk to pollinators and other beneficials. Some manufacturers have combined their spray products with insecticidal soaps to increase their effectiveness.

If you decide to utilize pyrethrins in your vegetable garden, use extreme caution. Read the label and follow the directions to the letter. Wear long sleeves, pants, and goggles to greatly reduce the chance of irritating your skin and eyes. Choose a calm day to eliminate pesticide drift.

To avoid exposing fish and other aquatic organisms to the pesticide, don't spray near water sources or storm drains, and do not pour any leftover solution down the drain!

Protect pollinators by applying pyrethrins in the evening or early morning, when they aren't active. Do not spray anywhere near flowering plants. Spot-spray the problem areas of your vegetable plants to minimize the amount of product you use.

While you can use pyrethrins up to the day of harvest, wait for the spray to dry before picking vegetables, and wash them thoroughly before eating them.

Pests Controlled by Pyrethrins

Pyrethrins control aphids, asparagus beetles, beet armyworms, blister beetles, cabbage loopers, cabbage worms, Colorado potato beetles, corn earworms, cucumber beetles, diamondback moths, earwigs, flea beetles, grasshoppers, harlequin bugs, Japanese beetles, leafminers, lygus bugs, Mexican bean beetles, onion maggots, spider mites, squash bugs, thrips, tomato fruitworms and hornworms, and whiteflies.

ROW COVERS

Row covers are my favorite and most useful organic control. If you use a physical barrier to cover plants that are susceptible to pests, the pests won't be able to access those plants to wreak havoc. I really prefer the idea of *excluding* pests from their host plants rather than trying to come up with ways to kill them.

The spun-bonded polyester fabric from which floating row covers are made allows sunlight and moisture to reach the plants underneath. It comes in different weights but the lightweight cover is most commonly used as a pest barrier. An added benefit of this weight is that it gives plants a few degrees of frost protection, which makes it useful as a season extender.

Some gardeners use heavier row covers called "frost blankets" to protect early- or late-season crops from getting frosted by cold temperatures. The downside to them is that they only allow 50 percent transmission of light, compared to 85 percent with lightweight floating row covers; less light transmission impacts plant growth. Because of this, the heavier weights are typically used on a short-term basis.

Whenever I'm growing cabbage family crops, I use bridal veil netting (see page 59) instead of floating row cover to protect them from aphids, cabbage loopers, and cabbage worms. Since these crops benefit from good air circulation, I've found the netting works better than regular row cover. It also allows me to more easily see what's going on under the cover without having to lift it and peer underneath. You can purchase tulle by the yard at fabric stores or by the bolt from online suppliers. Be sure to buy a very fine mesh with tiny holes because aphids are small and sneaky.

To apply floating row cover or netting over a crop, it is best to suspend it above the plants on a series of hoops or other supports. You can purchase premade hoops or make your own from flexible black poly sprinkler pipe or bent metal electrical conduit (refer to the DIY project on page 182). Once the row cover is in place, it's important to weight down the ends and sides to prevent it from blowing off on breezy days. After all,

A floating row cover is a versatile organic tool that acts as a barrier to prevent pest damage and provides frost protection for plants.

row covers are only effective if they remain over the crops you're trying to protect!

Keep in mind that you cannot leave row covers on vegetable plants for the entire growing season if they need to be pollinated in order to produce a crop. That doesn't mean you can't use them on those vegetables for part of the season. Some crops benefit from protection early in the season because the covers are an excellent way to prevent pests from laying their eggs and getting access to host plants for the early stages of their offspring's development.

If you've had problems in the past with animals, rodents, or birds bothering certain crops, there's an added benefit to using row cover. Covering a bed as soon as you plant seeds or seedlings makes them essentially invisible to those critters. Don't wait until *after* they've found the plants because they'll be determined to get to what they know is there. Again, if the crop will need to be pollinated, remove the cover as soon as the plants begin blooming.

Row covers will last several seasons if stored properly.

ORGANIC PEST MANAGEMENT PRODUCTS AND DIY PEST CONTROLS

SLUG AND SNAIL BAIT

Slugs and snails are frustrating garden pests because they are nocturnal. This means they are active and cause a lot of plant damage at a time when we can't see them and most predators aren't active. Using slug and snail bait is an easy way to get around those challenges; we leave little snacks for these mollusks to eat, they consume them, and then slither off to die.

Different baits have different formulations. Chemical baits contain metaldehyde, which is toxic to pets and other animals, so I do not recommend using it.

The active ingredient in organic slug and snail baits is iron phosphate, which naturally occurs in the soil and helps plants grow. It is safe for use around pets, wildlife, fish, birds, and beneficials, and can even be used on the day of harvest, although you should wash the produce thoroughly before consuming it. The bait comes in pellet form. Once slugs or snails consume even a small amount of it, they immediately stop feeding and die within a few days. Since they typically hide after eating the bait, you won't see dead bodies lying everywhere. You'll know it's working when the plant-nibbling dramatically decreases or stops.

Another form of organic slug and snail bait contains iron phosphate and spinosad, a naturally occurring soil bacterium. In addition to killing slugs and snails, this product will also kill earwigs, cutworms, pillbugs, and sowbugs. Applications last for 4 weeks. You can use both types of bait around ornamental plants, vegetables, berries, fruit trees, and in greenhouses.

To apply either of these organic baits, moisten the soil around the damaged plants to attract slugs and snails. Early evening is the best time to scatter the bait since these pests feed during the night and early morning hours. Slugs and snails exude a layer of mucus that helps them move about. Their slime trails will reveal the areas they travel, whether for feeding or hiding during the daylight hours. Their hiding places are great spots to treat.

Pick a day when there's no rain in the forecast for at least 24 hours. Keep in mind that snails and slugs are less active during especially dry or hot weather or when temperatures are quite cold. Avoid applying bait near water sources or storm drains.

In small areas, the application rate is 1 teaspoon of bait to 1 square yard of soil. Reapply the regular formulation of organic bait every 2 weeks, although you can apply it sooner if it's really being gobbled up. If you used the bait containing spinosad, the label directions indicate it shouldn't be applied more than 3 times in a 30-day period.

Don't touch your eyes if you've physically handled the product as it can cause irritation. Wash your hands well once you've finished the job, even if you wore gloves.

SPINOSAD

Spinosad is a biological insecticide derived from a fermented soil bacterium called *Saccharopolyspora spinosa*. It is effective against a range of damaging pests and widely used in organic agriculture. Spinosad kills both by contact and ingestion (consumption), with the latter method acting much more quickly. It attacks the nervous system, causing paralysis and then death. Pests typically die within 1 to 2 days.

A scientist first discovered the bacterium at an abandoned Caribbean rum distillery in the early 1980s. Spinosad is available in spray, dust, or granular form. Apply it to the upper and lower surfaces of the affected plants' leaves.

If you've already glanced at the sidebar showing which pests spinosad controls and think it is the answer to your garden challenges, you need to be aware of a very important issue.

While spinosad is safe to use around humans, mammals, and birds, it is highly toxic to bees and other pollinators. These insects are a critical part of our environment and, as such, gardeners and farmers need to take steps to protect them:

1. Try hand-picking the pests first to see if it's a viable method for eradicating them.

2. Only use a biological insecticide when you have a pest problem and the product label indicates it will control them.

3. Follow the directions on the label and remember that more is *not* better.

4. Never spray flowers.

5. Don't spray when pollinators are active. Apply it in the evening or early morning before pollinators are out and about. Allow spinosad to dry for a minimum of 3 hours. Research has shown that spinosad is primarily toxic to bees while it is still wet.

6. Avoid spraying near the host plants of butterflies and caterpillars.

7. Spray when it's calm. Windy conditions will cause the insecticide to drift to other areas.

8. Spinosad is slightly toxic to fish and other aquatic creatures. Do not apply it near water features.

9. Resistance is another important issue relating to the use of spinosad. When you use products containing this active ingredient over and over, pests can develop a resistance to it. Alternate the organic control methods you use to decrease the chance of this happening.

10. Refer to the product label for important information such as the maximum number of times you can apply it within a season, how long to wait before reapplying it, and when it is safe to harvest and consume sprayed produce.

Pests Controlled by Spinosad

Spinosad controls insects within the following orders: Lepidoptera (butterflies and moths), Diptera (true flies), and Thysanoptera (thrips). Applications will kill asparagus beetles, beet armyworms, cabbage loopers, cabbage worms, Colorado potato beetles, cucumber beetles, diamondback moths, flea beetles, harlequin bugs, leafminers, Mexican bean beetles, pillbugs, sowbugs, spider mites, thrips, tomato fruitworms, and tomato hornworms.

BEER TRAPS FOR SLUG CONTROL

Slugs are one of the most damaging and frustrating garden pests. It's bad enough when they make the leaves of our hosta plants look like Swiss cheese, but when they nibble on our prized vegetable plants, that is just not acceptable. In my garden, they seem to prefer salad greens and cabbage family crops. And the most annoying part is that they do it under the cover of darkness.

There are several methods for controlling slugs but creating a beer trap is one of the easiest. That's right, you can catch them using beer! It turns out slugs are attracted to the scent of yeast. Once they smell it, they'll head in that direction to investigate, lean over the container of beer to slurp some up, fall in, and drown. I guess there are worse ways to go, right?

My husband looks upon beer purchases with disdain when they're strictly for catching slugs. But unless you have a horrendous slug problem, there should be plenty left over to share with your favorite humans.

SUPPLIES

> Trowel

> Empty tuna or cat food cans, yogurt cups, or similar containers

> Beer of your choice (if it's only for the slugs, go with the cheap stuff since they aren't exactly connoisseurs)

Once the trap is in place, with the container's lip at the soil surface, add beer.

Success!

STEPS

1. Dig a hole with a trowel where the slugs are chewing on your veggies. Make it the size of your container and just deep enough that the lip of the trap is even with the soil. Repeat a few times to offer plenty of temptation.

2. Pour about ¾ inch (2 cm) of beer into each container. Do not fill your containers to the rim: If slugs barely have to reach for a sip, they'll slither away when they've had their fill.

3. Check your beer traps each morning. With a little luck, you'll find slugs blissfully floating in each one! Be aware that some slugs might be able to get back out of the container before you arrive on the scene. If this happens a lot, consider making the funnel beer trap (below).

Note: During hot weather, you will need to refill the cans more frequently due to evaporation.

FUNNEL BEER TRAP

SUPPLIES

- Empty plastic soda pop bottle (1- or 2-liter)
- Scissors
- Adhesive tape (masking, clear plastic, or duct tape)
- Trowel
- Beer (your choice)

STEPS

1. Use the scissors to cut off the top 4 inches (10 cm) of the bottle to make a funnel.

2. Trim down the lower portion of the bottle to about 5½ inches (14 cm) tall.

3. Cut a 1-inch-long (2.5 cm) slit into the cut edge of the funnel and another directly across from it. This helps the funnel fit firmly into the lower portion of the bottle.

4. Place the funnel spout-side down into the bottom section of the bottle. Tape the upper edge if it is quite sharp.

5. Dig a hole in the ground the same size as the trap, deep enough so the container's rim is at soil level.

6. Pour 2 inches (5 cm) of beer into the bottom of the bottle.

7. Check your traps each morning for slugs that died happy. Evaporation during excessively hot weather should be less of a problem than with the conventional trap, but do monitor beer levels.

1

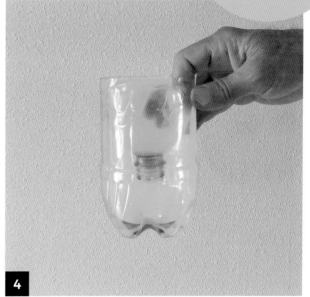

3

4

6

CARROT RUST FLY SCREEN

The larvae of carrot rust flies ruin carrots and parsnips by tunneling through the roots and filling the holes and grooves with rust-colored excrement. That doesn't sound very appetizing, does it? The adult flies find carrot plantings by the scent given off from their foliage and roots, and lay their eggs near the base of the plants. (Learn more about the life cycle of carrot rust flies in their profile on page 72.)

The traditional method for planting carrots and parsnips is to sow seeds thickly and thin them to a proper spacing later. Unfortunately, thinning plants releases the fly-attracting scent. By spacing the seeds at planting time, you won't have to handle plants until it's time to harvest them.

I love it when we can take advantage of a pest's limitations. In the case of carrot rust flies, these low fliers rarely get above the height of the plant foliage. All it takes is erecting a 30-inch-tall (76 cm) barrier around the bed to prevent them from gaining access to the plants.

In the Kitchen Garden at London's Kew Gardens, this screen protects carrots from carrot rust flies.

For this project, surround the planting with either floating row cover (page 145) or environmental mesh fabric (see Resources on page 190). I particularly like that the negligible (to us) height of the barrier allows for good air circulation and lets me easily peek over the top to check on plants. Be sure to put the barrier in place as soon as you plant the seeds to get ahead of any egg laying.

SUPPLIES

> Tape measure
> Hammer
> 40"-long (102 cm) wooden stakes—enough for the entire planting perimeter at 2'–3' (61–91 cm) spacing

> 36"-wide (91 cm) floating row cover OR environmental mesh fabric (long enough to surround the perimeter of your planting bed)
> Staple gun
> Staples

STEPS

1. Plant your seeds.

2. Pound in stakes around the perimeter of the planting bed to a height of about 31 inches (79 cm) above the soil surface, spacing them 2 to 3 feet (61 to 91 cm) apart. For this 3- by 8-foot (91 by 244 cm) raised bed, I used a total of 8 stakes.

3. Place the floating row cover or environmental mesh fabric along the outside of the stakes with a few inches resting on the ground; staple in place near the top and midpoint of each stake. Do not leave any gaps between the beginning and end of the row cover or mesh, because those flies will take advantage of them!

4. Cover the outer bottom edge of the barrier with soil so the flies can't gain access there.

5. At the end of the growing season, remove staples carefully to avoid tearing the row cover or mesh so you can use it again next season.

Note

Another way to protect your carrot family crops is by placing hoops and floating row cover over the bed for the entire season. The cover creates a physical barrier that keeps carrot rust flies away from the plants, provided you anchor the lower edge of it so the flies can't sneak in that way. The only disadvantage of this method is that you will need to occasionally lift the cover to see how the plants are coming along.

COPPER TAPE FOR SLUG CONTROL

Slugs are damaging pests that love to chew both ornamental and vegetable plants. They particularly seem to gravitate toward lettuce and cabbage family crops in my garden. On page 146, I explained how to use organic slug bait, and you'll find a DIY project for making beer traps on page 148. But I have one more trick up my sleeve, which involves the use of copper tape.

Did you know that the slimy skin of slugs reacts electrically to copper? This means copper makes a good repellent. Garden centers sell rolls of paper-backed copper tape you can affix to the outer perimeter of pots or other containers to create a barrier slugs won't cross. Rolls of copper foil adhesive tape are also available online for the same purpose.

The first time I used the paper-backed copper tape, I left the paper on it and made 3-inch-diameter (7.6 cm) rings held together with two staples. Then I placed them around the base of tender seedlings such as broccoli. The rings worked as repellents but as soon as they got wet, they flopped over and no longer made good contact with the soil, allowing slugs to gain access easily to the plants. So much for my keep-it-simple strategy.

Copper ring supplies

My husband, Bill, thought of an ingenious way to make sturdy rings that will last for years, so that's what we'll make here. They are ideal for plants with a single stem, such as broccoli, Brussels sprouts, kale, peppers, and tomatoes.

SUPPLIES

> 1 roll of copper tape barrier or copper foil adhesive tape

> 3"-diameter (7.6 cm) plastic drain pipe (10', or 305 cm, is the standard length)

> Handsaw

> Scissors

The copper rings do a great job of protecting broccoli.

STEPS

1. Cut the plastic pipe into 1½-inch (3.8 cm) sections with a handsaw.

2. Remove the paper backing from a section of tape while simultaneously adhering the tape to the outside of a ring. Repeat this step for each ring you will use.

3. At planting time, surround the base of each seedling with a ring, taking care that it makes good contact with the soil surface. Slugs are sneaky: If they spot an opening, they will go for it!

4. As the plants grow, make sure the leaves don't touch the ground outside the ring and create a ramp that circumvents the copper barrier.

5. At the end of the growing season, remove the copper rings, wipe off the soil, and store them in a dry location so they'll be in good shape for the next garden season.

Note

Use organic slug bait or beer traps when growing multi-stemmed or leafy crops, such as arugula, beets, cabbage, lettuce, spinach, and Swiss chard. You can surround the outer perimeter of a raised bed with copper tape, but if there are already slugs or their eggs in the soil, the plants you grow will still be at risk.

CUCUMBER BEETLE TRAP

Cucumber beetles are extremely damaging in the vegetable garden. In addition to munching on cucurbit family crops (cucumbers, melons, pumpkins, and summer and winter squash), these pests will also feast on asparagus, beans, beets, corn, potatoes, and tomatoes. As if that wasn't bad enough, they transmit two diseases, bacterial wilt and cucumber mosaic virus.

The adult beetles chew on the flowers, fruits, and stems of any of the previously mentioned vegetable plants. After the females lay eggs in the soil around the base of their preferred crops, the larvae hatch and feed on the plants' roots. By trapping the adults, you can disrupt their life cycle. Learn more about cucumber beetles in the pest profile on page 78.

This project takes advantage of the fact that the color yellow and the scent of clove oil, which mimics the odor of their pheromones (mating hormones), attract the beetles. The trap is simple and quick to make and doesn't require any special tools. The project as described here makes one trap.

Cucumber beetles stuck to a trap

SUPPLIES

- ❯ (1) 16-ounce yellow plastic drinking cup (the plastic makes the trap waterproof)
- ❯ Hole punch
- ❯ 2 twist-ties (from bread bags or grocery produce bags)
- ❯ 1 bamboo garden stake the height of your mature crop
- ❯ Sticky coating for insects (available at garden centers and online) or petroleum jelly
- ❯ Disposable gloves
- ❯ Wooden craft sticks, optional
- ❯ Cotton ball or round cotton pad
- ❯ Clove oil
- ❯ Eyedropper

STEPS

1. Use the hole punch to make 2 holes just below the lip of the cup, spaced about ½ inch (1.3 cm) apart.

2. Secure the cup to the stake with a twist-tie for each hole, as shown.

3. Put on your disposable gloves.

4. Spread a small amount of the sticky coating or petroleum jelly onto the inside bottom of the cup. The craft stick is helpful for this.

5. Place the cotton ball or cotton pad onto the sticky coating or petroleum jelly in the cup bottom.

6. Hold the stake and coat both the inside and outside of the cup with more sticky coating or petroleum jelly. If you get any of the sticky coating on your skin, use mineral oil, baby oil, or waterless hand soap to remove it.

7. Use the eyedropper to apply a few drops of clove oil to the cotton inside the cup.

8. Push the stake into the soil so the trap will be directly above the crop you are trying to protect. I used a 3-foot-long (91 cm) stake but you might need a taller one, depending on the height of your crop.

9. Monitor the trap frequently. As long as you are trapping cucumber beetles, refresh the sticky coating or replace the trap every 2 to 4 weeks. If it catches too many beneficials, remove it from the garden.

CUTWORM COLLARS

Cutworms love nothing better than chewing on the tender stems of young seedlings. Their favorites appear to be members of the cabbage and cucurbit families. It is so frustrating to carefully plant the seedlings you started indoors—or spent good money for at the garden center—only to discover several of them chopped off at ground level the next morning.

These annoying worms are nocturnal and spend their time either on the surface of the soil or just below it. They're often a problem in new gardens that were previously covered with grass. With this information in mind, you can protect your seedlings from the moment you plant them in the garden. Your goal is to keep the plants safe for the first couple of weeks while their stems are so tender, then they should be fine.

You'll notice I've suggested using paper products for the cutworm collars. This is because they will biodegrade. While it's true this will happen fairly quickly once they get quite wet, you only need to protect the seedlings for a short time. Some gardeners prefer to make reusable rings out of plastic drinking cups. These are wider and sturdier but you will need to remove them from the garden later. Choose the material that works best for you.

Empty toilet paper and paper towel rolls make excellent cutworm collars.

SUPPLIES

> Empty cardboard toilet paper or paper towel rolls or plastic drinking cups
> Newspaper strips 2½" (6.4 cm) wide, optional
> Measuring tape or ruler
> Marking pen
> Scissors

Note

Cutworm collars must be 1 inch (2.5 cm) above the soil surface and 1½ inches (3.8 cm) below. This is very important for protecting the plant!

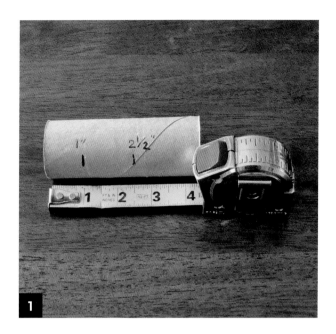

STEPS

1. Trim a toilet paper roll, paper towel roll, or plastic cup to a 2½ inch long (6.4 cm) ring. Be sure to remove the bottom of a cup. If your seedlings have root systems larger in diameter than your rings, you can either cut newspaper strips to size and form rings (use a few for each ring to make it sturdy) or use larger-diameter drinking cups.

2. Dig a hole that is deep enough to accommodate the seedling's rootball. Carefully slide one of the rings you made in step 1 over the seedling so it covers the lowest 1 inch (2.5 cm) of stem and 1½ inches (3.8 cm) of the root system.

3. Plant the seedling, making sure the roots of the plant make good contact with the soil and the collar stays in place. Press gently to remove any air pockets.

4. That's all there is to it. Check on your seedlings every day, just to be on the safe side. If you do notice some damage, use a trowel to carefully explore the soil around the seedling to locate the cutworm.

EARWIG TRAPS

When earwigs are ravaging your vegetable garden, it's hard to remember they're also considered beneficials because they prey on aphids, mites, and insect eggs. Some years, they experience population booms and cause a great deal of damage to plants. That was the case in 2019 when they chewed with great abandon on tender young seedlings, broccoli leaves, and other veggies in my garden. Many other gardeners reported the same problem. When this occurs, it's time to rein them in with simple-to-make traps.

Earwigs are nocturnal and hide under rocks, boards, and garden debris, or within dense leaves on plants during the day. By creating hiding places for them or luring them into traps, you can begin rounding them up. For this DIY project, I'll show you how to make two types of traps, one from cardboard or newspaper and one a container that holds a scent they won't be able to resist; it worked phenomenally well during the Earwig Plague of 2019!

While testing the cardboard traps, I made a surprising discovery: They also catch thrips. Since they are damaging insects as well (refer to page 118), this is a viable method for trapping them, too.

It's easy to catch earwigs in a simple trap containing vegetable oil and soy sauce.

CARDBOARD/NEWSPAPER TRAP SUPPLIES

> Corrugated cardboard with open ends OR sections of newspaper

> Twine

> Bucket

> Dish soap

STEPS

1. Create small rolls of cardboard as shown in the photo and secure each with twine.

2. Place your rolls onto the surface of the soil in a bed where earwigs have been damaging your plants. Alternatively, lay sections or rolls of newspaper onto beds with a lot of earwig activity. Leave the cardboard or newspaper out overnight.

3. In the morning, pick up the traps and shake the earwigs out of them into a pail of soapy water (use a few squirts of liquid dish soap to accomplish this). The soap will coat their bodies, making it difficult for them to escape.

4. As long as the cardboard or newspaper traps remain in good shape, place them back onto the garden beds as needed and repeat the morning emptying routine.

COVERED TRAP SUPPLIES

> Containers with tightly fitting lids, such as those from yogurt or buttery spread

> Inexpensive vegetable oil

> Soy sauce

> Tool for making holes in the lids (i.e., awl, large nail, small knife, cordless drill with a ¼" (6 mm) bit)

> Trowel

STEPS

1. Punch a few holes into the lids of your containers. The holes must be wide enough for an earwig to fit through, about ¼ inch (6 mm). You can also punch some holes into the containers themselves, about ½ to 1 inch (1.3 to 2.5 cm) below the top.

2. Dig a hole in the planting bed that will accommodate a container with its lip just above the soil surface.

3. Place the container into the hole.

4. Pour vegetable oil to 1 inch (2.5 cm) below the top of the container. Add a splash of soy sauce because the scent will attract earwigs. Snap the lid onto the container.

5. Check the container in the morning to see if there are some earwigs inside. Once the trap has quite a few of them in it, dispose of the oil and insects, then refill the container as needed.

INSECT HOTEL

Put out the welcome mat for beneficials! When you build an insect hotel, you provide them with places to lay eggs and/or hibernate. You're also giving your family the opportunity to discover the fascinating life of bugs. After building a simple structure, the most enjoyable part of the process is gathering natural materials from your garden to fill them. To learn more about insect hotels, refer to "Attract Pollinators and Other Beneficials to Your Garden" section on page 27. This project is for a 14- by 16-inch (36 by 41 cm) structure.

SUPPLY LIST

- (1) 1 × 8" × 8' (2.5 × 20 × 244 cm) board (insect hotel box and shelves)
- (1) 1 × 10" × 2' (2.5 × 25 × 61 cm) board (roof)
- (1) 2'-square (61 cm2) sheet of ½" (1.3 cm) plywood (back)
- (1) 2 × 4 × 7¼" (5 × 10 × 18 cm) board (center support for roof)
- 2 standard asphalt roofing shingles or other roofing material, as desired
- Box of 1⅝" (4 cm) wood screws
- Mason bee tubes (optional)

TOOLS

- Drill
- ⅛" (3 mm) drill bit
- Circular saw
- Utility knife

HABITAT FILLER MATERIALS

- Bamboo tubes
- Branches (small diameter) cut into 7" (18 cm) lengths
- Burlap
- Cardboard (corrugated), rolled or stacked
- Clay pot on side or inverted
- Driftwood
- Evergreen branches
- Hollow plant stems
- Logs drilled with ⁵⁄₁₆ × 12" (0.8 × 30.5 cm) holes
- Masonry bricks with holes
- Moss
- Pine needles
- Pinecones
- Reeds or straws
- Roofing tiles
- Straw
- Tree bark (loose sheets)
- Wooden blocks drilled with ⅝ × 6" (1.6 × 15 cm) holes

2

3

4

Note: For best results, always pre-drill your screw holes with the ⅛-inch (3 mm) drill bit before using the 1⅝-inch (4 cm) screws.

Make the insect hotel box first:

1. Cut the 1 × 8-inch × 8-foot (2.5 × 20 × 244 cm) board into (2) 14-inch (35.5 cm) pieces, (2) 16-inch (41 cm) pieces, and (2) 12½-inch (31.75 cm) pieces.

2. Use the ⅛-inch (3 mm) drill bit to pre-drill 4 holes on each end of the (2) 14-inch (35.5 cm) boards.

3. Place one of the 14-inch (35.5 cm) boards and one of the 16-inch (41 cm) boards at right angles to each other as shown and screw them together. Attach the other 16-inch (41 cm) board parallel to the first.

4. Put the other 14-inch (35.5 cm) board in place and screw it in place to finish the box frame.

5

6

5. Insert the 12½-inch (31.75 cm) shelves at 4 and 7 inches (10 and 18 cm) from the top of the box, respectively. Hold each shelf in place with screws through the side of the box and into the shelf ends.

Note: At this point, you can either proceed to step 6 to make a peaked roof or simply cover the top of the box with roofing shingles or another type of roof material and skip ahead to step 8.

6. To make the center roof support, place the 7¼-inch (18.4 cm) 2 x 4 at the center point of the top of the box (crosswise) and screw it into place from underneath.

7. Use a circular saw with the blade set at 30 degrees to cut the 1 × 10-inch × 2-foot (2.5 × 25 × 61 cm) board in half crosswise. When placed together, the beveled cuts will create the roof peak (flip one board over to accomplish this). Measure and trim the square end of each board 11 inches (28 cm) down from the roof peak. Place the boards flush with the back edge of the box, as shown in photo #8, to provide an overhang over the front of the board for a little extra weather protection. Screw the boards into place.

8. Laying the hotel box on its back on top of the sheet of plywood, use a pencil to trace the outer shape of the box. Trim the plywood to fit. Screw the plywood backing onto the box.

9. Use a utility knife to trim roof shingles as needed to fit the wooden roof pieces. Attach with roofing nails or other fasteners.

10. Completely fill each compartment with any of the items listed above as habitat materials.

11. Place your new hotel in a south- or east-facing location, ideally on a platform rather than the ground, so it will get plenty of morning sunshine.

RAISED BED WITH A ROW COVER TOP

For this project, we'll build a simple 3 × 8-foot (91 × 244 cm) raised bed and a hinged cover designed to keep damaging pests away from susceptible vegetable crops that don't require pollination. This is mainly designed for a floating row cover top but bird netting is another option to protect lettuce plantings from nibbling critters. I suggest using this bed on a three-year crop rotation plan by growing cabbage family crops one year (to keep away aphids, cabbage loopers, cabbage worms, and diamondback moth larvae), beet family crops the next (to thwart leafminers), and perhaps onions in the third year (if onion maggots are a problem).

This covered bed will keep the bugs away!

RAISED BED SUPPLIES

- ❯ (3) 2 × 10" × 8' (5 × 25 × 244 cm) boards (fir, pine, larch, cedar, or redwood)
- ❯ (20) 10 × 3" (25 × 7.6 cm) deck screws
- ❯ Electric or cordless drill
- ❯ 9/64" (3.6 cm) drill bit
- ❯ Screwdriver bit
- ❯ Level
- ❯ Tape measure

RAISED BED STEPS

1. Cut (2) 3-foot-long (91 cm) boards from **one** of the 2 × 10" × 8' (5 × 25 × 244 cm) boards.

2. Use the 9⁄64-inch (3.6 mm) drill bit to pre-drill 5 holes along each short edge of the 3-foot-long (91 cm) boards.

3. On a level surface, screw those boards onto the ends of the 8-foot (244 cm) boards.

4. Measure diagonally from opposite corners to determine if the bed is square; make adjustments as necessary so the two measurements are the same (tapping with a hammer will accomplish this).

5. Place the raised bed in its permanent location and use a level to determine if the bed is level on the ends and sides. Make it as level as possible to avoid water and soil running over the sides when watering the bed.

COVER SUPPLIES

- (3) 2 × 2" × 8' (5 × 5 × 244 cm) boards (frame base and top bar)
- (2) 2 × 2" × 8' (5 × 5 × 244 cm) boards cut into
  - (2) 2 × 2" × 3' (5 × 5 × 91 cm) boards (frame base ends)
  - (2) 2 × 2 × 18" (5 × 5 × 45.7 cm) boards (upright supports)
  - (2) 2 × 2 × 8" (5 × 5 × 20 cm) boards (top bar supports)
- (8) 8 × 3" (20 × 7.6 cm) screws
- (4) 8 × 2" (20 × 5 cm) screws
- (5) 8 × 1½" (20 × 3.8 cm) screws
- (10) 8 × ⅝" (20 × 1.6 cm) screws
- (4) 4" (10 cm) 90° metal flat corner braces with screws
- (3) ¼ × ¾" × 8' (0.6 × 2 × 244 cm) lengths of wooden window screen molding
- (2) 3½" (9 cm) square hinges with screws
- (1) 3⅞" (10 cm) metal handle with screws
- 25' (7.62 m) of ¾" (2 cm) black poly sprinkler pipe
- (10) ¾" (2 cm) poly pipe insert plugs
- Electric or cordless drill
- ⅛" (3 mm) drill bit
- Screwdriver bit
- Nail gun or finish nails and hammer
- Wood glue
- 6' × 12' (1.8 × 3.7 cm) piece of floating row cover or bird netting
- (2) 1 × 34" (2.5 × 86 cm) nylon straps

1A

COVER STEPS

1. **Make the frame base.**

   1A. Place (2) 2 × 2-inch × 8-foot (5 × 5 × 244 cm) boards 36 inches (91 cm) apart from each other. Attach the 36-inch (91 cm) boards on each end to make a rectangular frame base.

   1B. Measure diagonally from opposite corners to determine if the frame base is square; make adjustments as necessary.

   1C. Screw the 4-inch (10 cm) metal flat corner braces to each corner as shown in photo. These braces will be on the bottom side of the finished cover.

   Note

   Prior to screwing all of the 2 × 2-inch (5 × 5 cm) seams together, pre-drill the screw holes with a ⅛-inch (3 mm) drill bit and apply a coat of wood glue to each joint. Glue and an 8 × 3-inch (20 × 7.6 cm) screw secure each 2 × 2-inch (5 × 5 cm) corner joint.

1C

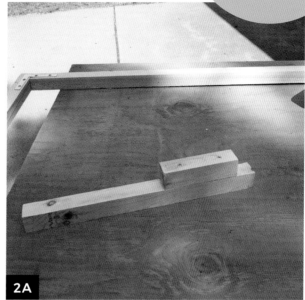

2A

2. **Assemble and install the (2) 18-inch (45.7 cm) uprights.**

2A. Pre-drill 2 holes into a 2 × 2-inch × 8-inch (5 × 5 × 20 cm) board and place it onto an 18-inch (45.7 cm) upright support 1½ inches (3.8 cm) from the top. Screw the 8-inch (20 cm) board in place with 2-inch (5 cm) screws. Repeat process with the other 18-inch (45.7 cm) upright.

2B. Mark the center (18 inches, 45.7 cm) on each 36-inch (91 cm) board of the frame base and pre-drill a hole from the top to the bottom. The flat corner braces from step 1c will be on the bottom of the cover.

2C. Use 3-inch (7.6 cm) screws and glue to attach uprights, with the top bar supports facing inward.

2C

Note

The 8-inch (20 cm) boards will provide support for the 2 × 2-inch × 8-foot (5 × 5 × 244 cm) top bar in step 3.

3. Rest the 2 × 2-inch × 8-foot (5 × 5 × 244 cm) top bar on upright supports and attach each end with glue and 3-inch (7.6 cm) screws.

4. **Attach poly pipe insert plugs to the frame base.**

   4A. Pre-drill a ⅛-inch (3 mm) hole through the bottom end of each insert plug.

   4B. To position the plugs, temporarily place a length of wood screen molding on top of the front edge of the 8-foot-long (244 cm) 2 × 2. Space 3 insert plugs about 2 feet (61 cm) apart. Screw the plugs into place with the 8 × ⅝-inch (20 × 1.6 cm) screws. (Note: If your plugs have tabs on them, make sure you position them parallel to the length of the boards.) Repeat along the back 8-foot-long (244 cm) 2 × 2.

   4C. Screw the 4 outer corner plugs in line with the others, securing them onto the 3-foot-long (91 cm) end boards, approximately ¾-inch (2 cm) in from

the edge. If your plugs have tabs on them, you will need to trim off the outer tab with a hacksaw or other tool.

5. **Attach hoops to the frame.**

   5A. Cut (5) 58-inch-long (147 cm) pieces of black poly sprinkler pipe.

   5B. Place the frame on the ground, then push the hoops onto the plugs.

TIP: *The easiest way to push each hoop onto a plug is to have a container of very hot water, dip the end of the pipe into it for about 30 seconds, then push the pipe onto the plug. Repeat the process until all 5 hoops are in place. Easy-peasy!*

   5C. Drill a ⅛-inch (3 mm) hole through the top center of each hoop down into the top bar. Use a 1 ½-inch (3.8 cm) screw to attach the hoops to it.

5B

5C

6A

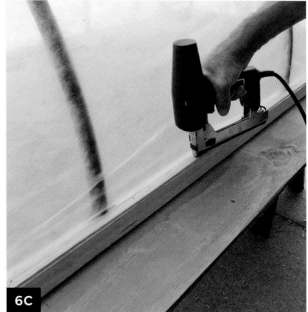

6C

6E

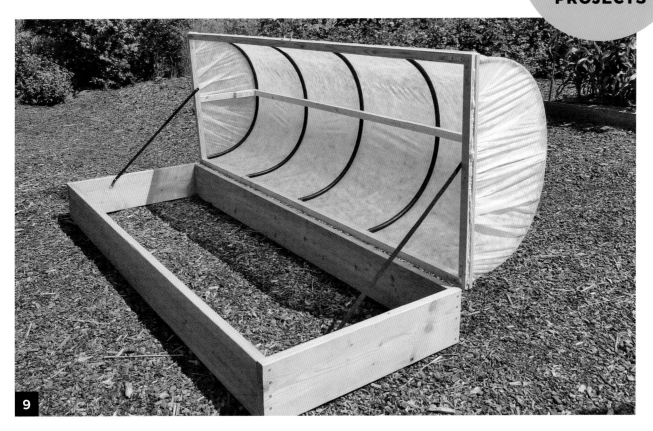

9

6. **Cover the frame.**

6A. Place floating row cover over the hoops with the excess draped evenly all the way around.

6B. Cut the window screen molding strips into

> (2) 5-foot (152 cm) pieces

> (2) 39¼-inch (99.7 cm) pieces

> (2) 3-foot (91 cm) pieces

6C. Working on the long sides first, place the 5-foot (152 cm) and 39¼-inch (99.7 cm) pieces lengthwise on top of the floating row cover, next to the insert plugs. Use a nail gun or finish nails about every 6 inches (15 cm) to secure them.

6D. Move to the opposite side of the frame. Carefully pull the floating row cover taut over the hoops. Place the 5-foot and 39¼-inch (152 and 99.7 cm) molding strips next to the plugs and nail down the strips.

6E. Moving to one end of the frame, neatly tuck together the excess row cover, place a 3-foot (91 cm) molding strip along the outer edge of the 2 × 2 and nail the strip into place. Repeat on the opposite end. Carefully trim the excess row cover all the way around.

7. Place the completed cover onto the raised bed. Attach with two hinges 24 inches (61 cm) from each end on the back of the cover and bed.

8. Install the metal handle at the front center of the cover.

9. To ensure the cover won't open too far, screw a nylon strap on the inside of one end of the raised bed approximately 10½ inches (26.7 cm) from the front. Screw the other end of the strap to the inside of cover at 13½ inches (34 cm) from the front, as shown. Repeat on the opposite end.

REFLECTIVE PLASTIC MULCH

Young vegetable seedlings attacked by pests either succumb to the damage or struggle to grow. If we protect them while they're young and tender, the plants will be vigorous and much more likely to survive future pest attacks. Reflective plastic mulch keeps some flying pests away from their favorite crops long enough to allow the plants to become established. The reflected light disorients winged aphids, Colorado potato beetles, flea beetles, leafhoppers, Mexican bean beetles, squash bugs, thrips, and whiteflies, making it difficult for them to find the plants. In addition to damaging vegetable crops, some of these pests are also vectors for disease—giving us extra incentive for outsmarting them.

Silver polyethylene mulch essentially "hides" the plants from these troublesome pests. It comes in 4 foot-wide (122 cm) rolls or sheets, and is sold online (see Resources on page 192). When you cover a garden bed with this mulch and plant seeds or seedlings through holes cut into it, the mulch surrounds the plants with a shiny glare. The mulch loses its effectiveness once the plant foliage covers more than 50 percent of the sheet. By that time, the plants should be flourishing.

In additional to controlling pests, this mulch increases the plants' productivity because it reflects more light into the foliage; cools the soil, making it particularly ideal for cool-season crops; and reduces weeds.

The reflective nature of silver mulch makes it difficult for some pests to find their host plants.

SUPPLY LIST

> Silver polyethylene mulch
> Landscape staples to hold down mulch* (optional)
> Scissors

> Trowel
> Bucket

STEPS

1. If you have a drip irrigation system or soaker hose for the bed, put it in place first.

2. Moisten the soil so the silver polyethylene mulch will make good contact with it.

3. Cut the mulch to the appropriate size for your planting bed.

Protect These Vegetables with Reflective Mulch

If aphids, Colorado potato beetles, flea beetles, leafhoppers, Mexican bean beetles, squash bugs, thrips, or whiteflies have been a problem in your garden or region, plant bush beans, large cabbage family crops (broccoli, Brussels sprouts, cabbage, and cauliflower), cucurbit family crops (cucumbers, melons, pumpkins, summer and winter squash), and nightshade family crops (peppers, potatoes, and tomatoes) in beds covered with silver polyethylene mulch. All of these crops are well-suited because they are spaced at wide intervals.

4

4. Dig a small trench around the perimeter of the bed (place the excess soil in a bucket) and weight down the outer 3 inches (7.6 cm) of the mulch with that soil or pin down the mulch with the landscape staples*. Both of these methods ensure the mulch remains in good contact with the soil and won't blow off.

5. Plant each seed or seedling by cutting a small hole into the mulch at appropriate intervals and then planting through each hole.

*Note

Make your own landscape staples or purchase them online. The more holes in the silver mulch, the more fragile it will become. This impacts your ability to reuse the mulch in a subsequent growing season. With a bit of care, it is possible to remove the mulch and use it again.

ROW COVER HOOPS

Row covers (see page 145) are an effective organic tool for controlling many pests. They work best when suspended above vegetable plants on hoops. In this DIY, you'll learn how to make hoops from two different materials: ½-inch-diameter (1.3 cm) electrical metal tube (EMT, which is also called electrical conduit or metal conduit) OR ¾-inch-diameter (2 cm) black poly sprinkler pipe.

EMT is extremely sturdy so it will last for years and even stands up to a snow load if you want to grow vegetables under a plastic-covered low tunnel during the winter. Black poly sprinkler pipe is inexpensive (you can make about 15 hoops from a 100-foot [30.5 m] roll), doesn't require any special equipment to bend, is very durable, and lasts for decades; it won't hold up to a snow load, however. While you can push the EMT hoops into the soil, it's easiest to support the poly pipe hoops by slipping them over metal stakes. In my garden, I space our hoops 2 to 3 feet (61 to 91 cm) apart.

Sticks of EMT make very sturdy hoops for row covers.

SUPPLIES

> Sticks of ½"-diameter (1.3 cm) EMT (sold in 10' (25 cm) lengths at hardware stores)
> (1) 2 × 4' (61 x 122 cm) sheet of ¾" (2 cm) plywood or chipboard
> (2) 2 × 2" × 8' (5 × 5 × 244 cm) boards cut into
>> (23) 2 × 2 × 4" (5 × 10 cm) boards
>> (2) 2 × 2 × 8" (5 × 5 × 20 cm) boards
> (1) 1 × 2" × 2' (2.5 × 5 × 61 cm) board

> Cordless drill
> Drill bits
> Screwdriver bit
> (60) 2" (5 cm) screws
> Tubing cutter or hacksaw
> Tape measure
> Pencil

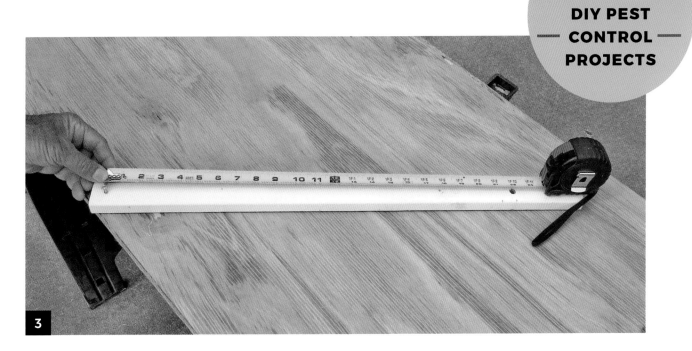

3

EMT HOOPS

For this part of the project, we'll make a bender to create the 44-inch-wide (112 cm) EMT hoops. If you only need a few hoops, this homemade bender is inexpensive and easy to use. If you need a lot of hoops, consider purchasing a commercial pipe bender from a home center or online source. You could even go in on the purchase with other gardening friends.

1. For each stick of EMT, make a mark at 14 inches (36 cm) and at 97 inches (246 cm).

2. Pre-drill 2 screw holes into each 4-inch (10 cm) piece of 2 × 2 using a ⅛-inch (3 mm) drill bit; pre-drill 4 screw holes into each 8-inch (20 cm) piece; you will attach these to the plywood later.

3. Drill a ⅛-inch (3 mm) pilot hole ¼ inch (6 mm) from an end of the 1 × 2-inch × 2-foot (2.5 × 5 × 61 cm) board. Measure 22 inches (56 cm) from that hole and drill a ⁵⁄₁₆-inch (8 mm) hole through the board.

Calculating EMT Hoop Size

1. Measure the width of the soil you want the hoop to span. Divide that number in half to find measurement A, which is the radius of a circle.

2. Multiply (A) by 3.14 (pi) to get measurement B, the length of EMT you will need so far.

We'll use a 44-inch-wide (112 cm) inside measurement of a bed as an example.

Divide the width of the bed (44"/112 cm) by 2. This gets you a 22" (56 cm) radius, so measurement A is 22" (56 cm).

Multiply 22" (56 cm) by 3.14 (pi) to get 69.08" (175 cm)—we'll round down to 69" (175 cm). This is the length of EMT needed to span the bed.

Because the center of our hoop is only 22" (56 cm), we'll make it taller by adding a 14" (36 cm) leg to each side. The total length of EMT we need to make our hoop is 97" (246 cm), because 69" + 14" + 14" = 97" (175 cm + 36 cm + 36 cm = 246 cm).

Note: If you want the hoop to be higher than measurement (A), determine how tall you want each "leg" of the hoop to be (C) and add that number twice to (B). (B) + (C) + (C) = the total length of the EMT you need to make a hoop that size. Remember that sticks of EMT are 10 feet (3 m) long, so that is the maximum length you can use.

4B

4C

4. Create an arc guideline for the hoop.

4A. Place the sheet of plywood onto a work surface. Measuring from the left side, make a mark at 24½ inches (62 cm) along the front edge. Set the 1 × 2-inch × 2-foot (2.5 × 5 × 61 cm) board on top of the plywood with a screw in the ⅛-inch (3 mm) pilot hole. Align the board so the screw is about ¼ inch (6 mm) above the 24½-inch (62 cm) mark and screw it to the plywood.

4B. Holding a pencil upright in the ⁵⁄₁₆-inch (8 mm) hole, pivot the 1 × 2 over the plywood to draw an arc. Unscrew and remove the 1 × 2 board.

4C. From the left side, begin attaching the 4-inch-long (10 cm) 2 × 2 pieces using 2-inch (5 cm) screws. Space them evenly along the inside of the arc. Use a spare 2 × 2 as a spacer. At the inside right end of the arc, screw an 8-inch-long (20 cm) 2 × 2 with the long side parallel and flush with the front of the plywood. Approximate the spacing of the last 2 or 3 blocks on the right side to complete the arc support.

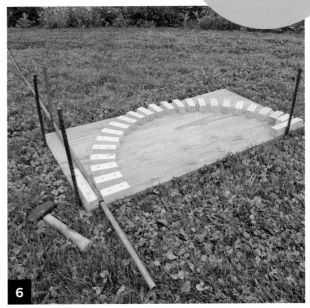

5. On the left side, temporarily use an EMT stick as a spacer and install the remaining 8-inch-long (20 cm) 2 × 2 on the left edge of the plywood, perpendicular to the front. You want the space to be snug but easily accommodate the width of an EMT stick when you start bending them.

6. Move the plywood bender to a flat work area, such as a lawn, or attach it to a sturdy bench. If positioning on the ground, pound stakes in the locations shown to stabilize the board against the torque generated while bending the EMT. Slip the end of the first EMT into the left side of the bender, making sure the 14-inch (36 cm) mark (leg measurement) is even with the front edge of the plywood.

7. While firmly holding the far end of the EMT, keep it low and parallel to the sheet of plywood and slowly pull/bend it around the form.

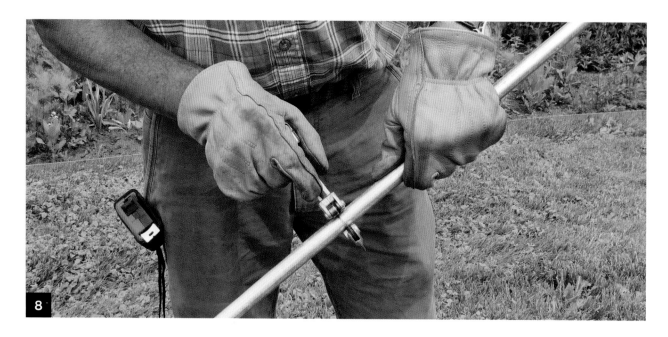

8. Remove the hoop from the bender. Use a tubing cutter or hacksaw to trim off the excess at the 97-inch (246 cm) mark. Measure 14 inches (36 cm) from the cut end and mark the tube.

9. Flip around the end of the hoop you just cut so the mark from the cut end is even with the front left of the hoop bender. Re-bend the hoop to make it more symmetrical. Remove it from the bender.

10. Since EMT wants to be straight, you might need to stand the hoop in a "C" shape and apply a little pressure to the top leg to make sure it's the desired width for your garden bed.

BLACK POLY PIPE HOOPS

The following steps will guide you through the process for making custom hoops.

SUPPLIES

> 100-foot-long (30.5 m) roll of ¾"-diameter (2 cm) black poly sprinkler pipe

> Hacksaw or poly pipe cutter

> Tape measure

> 1'-long, ⅜"-diameter (30.5 cm long, 1 cm diameter) rebar stakes (2 per hoop)

STEPS

1. Calculate how large you would like your hoops to be by unrolling a portion of the poly pipe and holding it in place over a bed at your preferred height and width. Mark the poly pipe at that length. I used 72 to 75 inches (183 to 191 cm) of poly pipe for hoops that fit our 3- and 4-foot-wide (91-cm and 122-cm) raised beds, but if you will need to cover taller crops, such as broccoli or kale, you might require slightly longer hoops. Use the instructions in the Calculating EMT Hoop Size sidebar on page 183.

2. Place the roll of poly pipe in the sun for a while so it will lie flat. Use a tape measure to determine the length of the pipe you marked, then mark enough sections of pipe for the number of hoops you need.

3. Use a hacksaw or poly pipe cutter to cut the pipe into the hoop lengths.

4. To support the hoops, pound rebar stakes halfway into the soil on either side of the bed at your desired spacing and slip your new poly pipe hoops over them.

4

STICKY TRAPS

Sticky traps have two useful purposes in the garden: They indicate if certain insects are present and they trap pests. In the first instance, you can implement control methods if you know when an pest has arrived. Many commercial greenhouses use sticky traps to watch for whiteflies among the plants they're growing. In my small orchard, I hang sticky traps to learn when apple codling moths and cherry fruit flies become active so I can begin our organic spray routine. Vegetable gardeners often use the traps to monitor for damaging pests, such as flea beetles.

When used specifically to kill garden pests, sticky traps help control winged aphids, leafhoppers, leafminers, Mexican bean beetles, moths, and thrips.

The concept behind the traps is this: Choose a waterproof material in a color that is known to attract the damaging pest, apply a sticky coating for insects (found online or in garden centers) or petroleum jelly, and hang the trap near the target crop. The pests head to the trap, get stuck, and die.

The traps might catch parasitic wasps or beetles so it's very important to monitor them regularly and be ready to remove them if they are catching beneficials. You'll find a link to a helpful guide on the pests most commonly found on sticky traps in the Resources section on page 194. It's interesting to note that both thrips and whiteflies turn orange when caught in the adhesive.

While commercial sticky traps are available at garden centers and online, you can easily make your own. With the exception of thrips, which are more drawn to the color blue, the rest of the insects listed are attracted to yellow.

Leafhoppers (and a fly or two) found this sticky trap irresistible.

SUPPLIES

> Yellow or blue plastic picnic plates, plastic drinking cups, sturdy plastic folders, or plastic-coated cards

> Sticky coating for insects or petroleum jelly

> Wooden craft stick or single-use paint brush

> Disposable gloves (optional)

> Hole punch

> Wire for hanging traps

> Scissors

STEPS

1. If using plastic folders or cards, cut them into a manageable size, such as 3 by 8 inches (7.6 by 20 cm).

2. Punch a hole about ½ inch (1.3 cm) from one edge of your trap base. Insert a piece of wire so the trap is ready to hang as soon as you've coated it in the next step.

3. Use a brush or a wooden craft stick to coat the surface of the trap with either sticky coating or petroleum jelly. You can coat both sides of the trap if you are hanging it from above without the risk of it sticking to a plant or support. Note: If you're using a sticky coating, it is inevitable that you'll get some on your hands. Use mineral oil, baby oil, or waterless hand soap to remove it—or wear disposable gloves for this step!

4. Carefully hang the trap directly above, but not touching, the plants you are trying to protect. You will likely need to adjust the height as the plants grow.

Note

Are cucumber beetles a problem in your garden? There's a DIY trap for them on page 156.

RESOURCES

The volume of information available online, in books, and through local resources, such as garden centers and nurseries, can be overwhelming. No worries! This resource section lists some of my favorite books, podcasts, and websites to get you started. I've included a table of suppliers of beneficials, row covers, and other products I mention. You'll also find the bug mugshots here, a gallery of most and least wanted designed to help you put names to faces so you can roll out the welcome mat or use the tools in this book to say scram! But first, some basic resources.

BOOKS

Attracting Beneficial Bugs to Your Garden
Jessica Walliser, 2014

Attracting Native Pollinators
The Xerces Society, 2011

Farming with Native Beneficial Insects
The Xerces Society, 2014

Field Guide to Insects and Spiders and Related Species of North America
National Wildlife Federation, 2007

Garden Insects of North America, Second Edition
Whitney Cranshaw and David Shetlar, 2017

Gardening for Butterflies
The Xerces Society, 2016

Gardening for the Birds
George Adams, 2013

Good Bug Bad Bug
Jessica Walliser, 2011

The New Organic Grower, 3rd Edition
Eliot Coleman, 2018

Pollinator Friendly Gardening
Rhonda Fleming Hayes, 2016

The Pollinator Victory Garden
Kim Eierman, 2020

ORGANIC GARDENING PODCASTS

Back to My Garden
Dave Ledoux
backtomygarden.com

Green Organic Gardener
Jackie Marie Beyer
organicgardenerpodcast.com

The Grow Guide
Maggie Wysocki and Dave Hanson
thegrowguide.libsyn.com

The joe gardener Show with Joe Lamp'l
joegardener.com/podcasts

The Organic Gardening Podcast
Chris Collins and Sarah Brown
gardenorganic.org.uk/podcast

Permaculture & Ecology Podcast
Sustainable World Radio
sustainableworldradio.com

Make every trip into your garden a joyful and educational experience.

ONLINE HELP WITH IDENTIFICATION

Apps

Picture Insect – Bug Identifier

Insect Identification (Available for Apple only)

Leps by Fieldguide

Seek by iNaturalist

Websites

BugGuide.net (Iowa State University Department of Entomology)

InsectIdentification.org

www.KnowYourInsects.org

Facebook Groups: Look for Facebook groups that help identify insects in your region, such as the Pacific Northwest Bugs group. There are also national and international groups that identify insects, including the Insect Identification group; Bug Identification; Bug Identification Group – People Helping Others ID Bugs (note: this is different from the previous one listed); and the Insect, Arachnid, and Bird Appreciation and Identification group. To locate a list of groups, log onto Facebook and conduct a search using the words "insect identification."

How to Identify Insects on Sticky Traps: content.ces.ncsu.edu/insects-found-on-yellow-sticky-traps-in-the-greenhouse

EDUCATIONAL SOURCES

How to Locate Master Gardener Programs in the United States and Canada mastergardener.extension.org/contact-us/find-a-program/

IPM (Integrated Pest Management) Institute of North America

ipminstitute.org/what-is-integrated-pest-management/

Regional Insect and Gardening Information from United States Extension Programs

Because educational institutions occasionally change servers, Susan is maintaining an updated list of suggested Extension programs on her website: SusansintheGarden.com/guides/links.

United States Dept. of Agriculture Hardiness Zone map

planthardiness.ars.usda.gov/PHZMWeb/

BIRDWATCHING REFERENCES

National Audubon Society: audubon.org

Cornell Lab of Ornithology: allaboutbirds.org (this includes dimensions for birdhouses)

AUTHOR RESOURCES

SusansintheGarden.com: Blog; guides for growing vegetables, preserving the harvest, and birds in the garden; Susan's weekly garden columns; index to her videos; photo galleries

Susan's gardening videos: youtube.com/susansinthegarden

Note: We've made every effort to provide current web addresses for these sources, but businesses and educational institutions occasionally change servers or close. If a link doesn't work, please conduct a web search for the name given here. With a little luck, the results should steer you in the right direction.

PRODUCT SUPPLIERS

	Beneficials	Environmental mesh/screen	Organic pest control products	Insect hotels	Mason bees	Plastic mulch	Row cover	Traps
Arbico Organics 10831 N Mavinee Dr., Suite 185 Oro Valley, AZ 85737 800-827-2847 arbico-organics.com	X		X					X
Buglogical Control Systems PO Box 32046, Tucson, AZ 85751 520-298-4400 buglogical.com	X							
Crown Bees 13410 NE 177th Pl., Woodinville, WA 98072 crownbees.com					X			
Extremely Green Gardening Company PO Box 2021, Abington, MA 02351 781-878-1800 extremelygreen.com	X		X					
Fedco Seeds and Supplies PO Box 520, Clinton, ME 04927 207-426-9900 fedcoseeds.com						X	X	
Gardener's Edge 241 Fox Dr., Piqua, OH 45356 888-556-5676 gardenersedge.com	X	X	X				X	
Gardener's Supply 128 Intervale Rd., Burlington, VT 05401 800-876-5520 gardeners.com			X	X			X	
Gardens Alive PO Box 4028, Lawrenceburg, IN 47025 513-354-1482 gardensalive.com	X		X				X	X
Greenhouse Megastore 888-281-9337 Greenhousemegastore.com		X					X	
Gurney's Seed & Nursery Company PO Box 4178, Greendale, IN 47025 513-354-1491 gurneys.com			X				X	

	Beneficials	Environmental mesh/screen	Organic pest control products	Insect hotels	Mason bees	Plastic mulch	Row cover	Traps
Harris Seeds PO Box 24966, Rochester, NY 14624 800-544-7938 harrisseeds.com		X	X			X	X	X
Johnny's Selected Seeds PO Box 299, Waterville, ME 04903 877-564-6697 johnnyseeds.com			X			X	X	X
March Biological Control 800-328-9140 marchbiological.com	X							
Mason Bees for Sale 10090 N. Highway 38, Deweyville, UT 84309 801-648-9035 masonbeesforsale.com					X			
Mason Bees Supply Urban Pollinators, LLC Eugene, OR 541-654-5224 mason.bees.supply					X			
Natural Gardening Company PO Box 258, Diamond Springs, CA 95619 707-766-9303 naturalgardening.com						X	X	
Peaceful Valley Farm Supply 125 Clydesdale Ct., Grass Valley, CA 95945 888-784-1722 groworganic.com			X		X		X	X
Planet Natural Bozeman, MT 59715 888-349-0605 planetnatural.com	X		X					X
Safer Brand 855-767-4264 saferbrand.com			X					
Tanglefoot The Scotts Miracle-Gro Company 14111 Scottslawn Rd., Marysville, OH 43041 888-295-6902 tanglefoot.com			X					X
Territorial Seed Company PO Box 158, Cottage Grove, OR 97424 800-626-0866 territorialseed.com			X				X	X

BUG MUGSHOT GALLERY

When you come across a bug you don't recognize, use this handy gallery of photographs to match it with a name. Then you can head to the pest profiles beginning on page 60 or the beneficials table beginning on page 124 to learn about it and determine whether its presence is cause for alarm or celebration!

MUGSHOT	BUG	PAGE #
	Aphid	60
	Asparagus beetle, common	62
	Asparagus beetle, spotted	62
	Assassin bug	124

MUGSHOT	BUG	PAGE #
	Beet armyworm larvae	64
	Beet armyworm moth	64
	Big-eyed bug	124
	Blister beetle, margined	66

BUG MUGSHOT GALLERY

MUGSHOT	BUG	PAGE #
	Japanese beetle grubs	94
	Japanese beetle	94
	Lacewing, green	126
	Lacewing larva	126
	Ladybug	127

MUGSHOT	BUG	PAGE #
	Ladybug larva	127
	Leafhopper, beet	96
	Leafhopper, potato	96
	Leafminer fly	98
	Leafminer larva	98

MUGSHOT	BUG	PAGE #
	Robber fly	129
	Root maggot (onion maggot)	106
	Rove beetle	130
	Slug	108
	Snail	108

MUGSHOT	BUG	PAGE #
	Snakefly	130
	Soldier beetle	130
	Sowbug	104
	Spider	131
	Spider mite	110

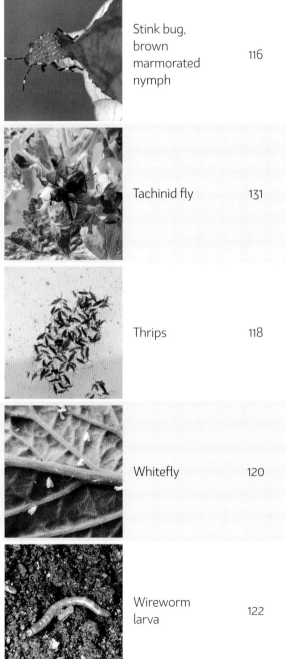

ABOUT THE AUTHOR

Susan Mulvihill began growing vegetables as a teenager while living in Southern California. Once she and her husband, Bill, moved to Spokane, Washington, in 1978, the gardening bug really took hold.

She has been sharing her passion for it with the public since 2000. At that time, Susan was writing articles for Spokane's largest garden club, the Inland Empire Gardeners. This quickly led to writing the weekly garden columns for *The Spokesman-Review*, where she has worked both as a newsroom employee and correspondent for more than 30 years.

In 2002, Susan became a Spokane County Master Gardener, which has provided her the opportunity to answer all manner of gardening questions for the public ever since. Her focus has always been on spreading the word that organic gardening is simple and effective.

As a longtime member of the professional organization Garden Communicators International, Susan has won awards both for her writing and for designing her popular blog, SusansintheGarden.com.

In 2014, Susan and her friend, Pat Munts, teamed up to write the *Northwest Gardener's Handbook* for Cool Springs Press. It quickly became a popular resource for folks gardening in the diverse climates and terrain of the northwest United States.

Susan received the 2019 Media Award from the Master Gardener Foundation of Washington State, being recognized for her "many contributions to the coverage of gardening topics, education, and demonstrations that emphasize the Washington State University Extension Master Gardener program."

Several years ago, Susan embraced the concept of video storytelling as a tool for teaching gardening. She has created more than 300 how-to-garden videos for her YouTube channel, youtube.com/susansinthegarden. They cover a wide variety of topics, although Susan primarily focuses on vegetable gardening.

She posts daily on her Facebook page (facebook.com/susansinthegarden) and Instagram feed (Instagram.com/susansinthegarden).

Susan and Bill enjoy gardening together on their five acres, although occasional, good-natured spats over who gets to plant what in the raised beds have been known to occur. In 2017, their garden was featured on the popular PBS series *Growing a Greener World* (episode 809). You can reach Susan via email at Susan@susansinthegarden.com.

ACKNOWLEDGMENTS

Writing a book during a pandemic has been an interesting experience for me. I had to cancel my speaking engagements and found I had much more time on my hands than usual. To my mind, the most positive impact of the pandemic has been the surge of interest in growing your own food. I have enjoyed answering new gardeners' questions and kept them in mind as I wrote each chapter. It's probably not surprising that this project also distracted me during such unsettling times.

Writing about insects was another benefit: It made me much more aware of what's been going on under my nose. Every trip into the garden became an adventure during which I was equally excited to see both the good and bad bugs out there. It's probably a safe bet that my reveling in the bad bugs won't continue into the next growing season!

A downside of the pandemic was not being able to travel. When you're writing a book about vegetable garden pests but you live in a region that doesn't get a lot of the nasty ones that plague so many gardeners, that certainly limits the opportunities to photograph them. In this case, it all came together in spite of the challenges.

This book would not have been possible if it weren't for two very special people: Jessica Walliser and Bill Mulvihill.

In August 2019, I pitched an idea for a book to Jessica, the acquisitions editor for Cool Springs Press' gardening titles. My initial plan was for an all-encompassing book on vegetable gardening, with heavy emphasis on organic practices. Jessica immediately saw the potential of focusing on organic pest control while incorporating the many aspects of successful gardening that I had wanted to share with readers. After a phone call or two, she and I were so excited about the concept that we couldn't wait to get started. Thank you, Jessica, for recognizing how useful this book will be for gardeners everywhere and for giving me this opportunity.

My amazing husband, Bill, has been through this book-writing routine before (I co-authored the *Northwest Gardener's Handbook* in 2014). He knows how time-consuming and stressful it can be to make everything come together, and probably wondered why I would take on a project of this magnitude. But I knew this book *needed* to happen. Bill read every single word, made suggestions, and let me bounce ideas off of him. He figured out the technical details for the more challenging DIY projects (I'm a writer, not a carpenter), helped me build them, and bolstered me with words of encouragement. I cannot thank you enough, honey.

Thanks also go to project manager Meredith Quinn, art director Regina Grenier, and the talented staff at Cool Springs Press. Their hard work and expertise have made *The Vegetable Garden Pest Handbook* look sharp and read smoothly.

I would be remiss if I didn't thank my gardening friends and Master Gardener colleagues. They welcomed me into their gardens to photograph some of the insects in this book and answered the occasional question that invariably began with the words "how have you dealt with . . . ?" This includes Steve and Marie Cole, Bruce Kaufman, Jonie Knoell, Bernie Labrucherie, Julie McElroy, Pat Munts, Lucy Potts, Marcia Sands, Jackie Sykes, JoAnn Townsend, Brianne Tucker, Sharon Watson, and David Yarbrough.

As always, the support of my gardening-loving family means the world to me. Thank you to my sisters Kathy, Lucy, and Anne, for always inspiring me and sharing my passion for growing things!

INDEX